Graduate Texts in Mathematics **101**

Springer
New York
Berlin
Heidelberg
Barcelona
Budapest
Hong Kong
London
Milan
Paris
Santa Clara
Singapore
Tokyo

Graduate Texts in Mathematics

continued after index

Harold M. Edwards

Galois Theory

 Springer

Harold M. Edwards
New York University
Courant Institute of
    Mathematical Sciences
251 Mercer Street
New York, NY 10012
USA

AMS Classifications: 12-01, 12-03, 12A55, 01A55

Library of Congress Cataloging in Publication Data
Edwards, Harold M.
  Galois theory.
  (Graduate texts in mathematics; 101)
  Bibliography: p.
  Includes index.
  1. Galois theory. I. Title. II. Series.
QA247.E383   1984   512´.32   83-20082

Printed on acid-free paper.

Typeset by Composition House Ltd., Salisbury, England.
Printed and bound by R. R. Donnelley & Sons, Harrisonburg, VA.
Printed in the United States of America.

9 8 7 6 5 4 3 (Corrected third printing, 1998)

ISBN 0-387-90980-X Springer-Verlag New York Berlin Heidelberg
ISBN 3-540-90980-X Springer-Verlag Berlin Heidelberg New York   SPIN 10644911

To Betty

Preface

This exposition of Galois theory was originally going to be Chapter 1 of the continuation of my book *Fermat's Last Theorem*, but it soon outgrew any reasonable bounds for an introductory chapter, and I decided to make it a separate book. However, this decision was prompted by more than just the length. Following the precepts of my sermon "Read the Masters!" [E2], I made the reading of Galois' original memoir a major part of my study of Galois theory, and I saw that the modern treatments of Galois theory lacked much of the simplicity and clarity of the original. Therefore I wanted to write about the theory in a way that would not only explain it, but explain it in terms close enough to Galois' own to make his memoir accessible to the reader, in the same way that I tried to make Riemann's memoir on the zeta function and Kummer's papers on Fermat's Last Theorem accessible in my earlier books, [E1] and [E3]. Clearly I could not do this within the confines of one expository chapter.

And so I decided to write a short book—a sort of volume $1\frac{1}{2}$ of my work on Fermat's Last Theorem—devoted entirely to the basics of Galois theory. There is very little in this book that is not already to be found, however concisely and however lacking in proof, in Galois. The one major exception is the material on factorization of polynomials (§§49–61), which is due to Kronecker and which seems to me to be necessary to give clear meaning to the computations with roots of algebraic equations that Galois and Lagrange performed without inhibition and without comment.

The crux of Galois theory is, appropriately enough, Galois' Proposition I, which is the following characterization of what we call the *Galois group* of an equation. Let $a, b, c, \ldots$ be the n roots (assumed distinct) of an algebraic equation $f(x) = 0$ of degree n. The Galois group is a certain subgroup of the group of permutations of the roots $a, b, c, \ldots$. Galois used it to determine whether a given polynomial in the roots $F(a, b, c, \ldots)$ has a known value—in modern parlance, to determine whether $F(a, b, c, \ldots)$ is in the ground field. The characteristic property of the Galois group is that $F(a, b, c, \ldots)$ has a known value if and only if

$$F(a, b, c, \ldots) = F(Sa, Sb, Sc, \ldots)$$

for all permutations S of the Galois group. Galois proved the existence and uniqueness of a group with this property by *constructing* it, using what later became known as a Galois resolvent. (This characterization of the Galois group will be more recognizable to readers familiar with modern formulations of Galois theory after they read the first corollary in §41. See also §63.)

The major theorems of Galois, such as the theorems on the solvability of equations by radicals, flow from the study of the relationship between algebraic equations $f(x) = 0$ and the groups associated with them. Of particular importance is the analysis of the way in which the group is reduced when the field of known quantities is extended (Galois' Propositions II–IV).

Some recent texts on Galois theory place mistaken emphasis on the question of finding explicit quintic equations, with rational coefficients, which cannot be solved by radicals. This is a moderately interesting result (one not covered in this book) but it is not a key theorem of Galois theory. Galois showed that an algebraic equation is solvable by radicals if and only if the associated group is solvable. A given quintic with rational coefficients can therefore be tested for solvability. Abel's theorem that the *general* quintic is not solvable states that the equation $x^5 + Bx^4 + Cx^3 + Dx^2 + Ex + F = 0$—an equation with coefficients in the field $\mathbb{Q}(B, C, D, E, F)$ obtained by adjoining five transcendental elements (variables) to $\mathbb{Q}$—is not solvable by radicals. (In Galois theory this follows from the fact that the Galois group of this equation is the full group of 120 permutations of the five roots.) In other words, no field extension of $\mathbb{Q}(B, C, D, E, F)$ obtained by a succession of adjunctions of radicals can ever contain a root of the given equation. This is what it means to say that the quadratic formula

$$x = \frac{-B \pm \sqrt{B^2 - 4C}}{2},$$

and the much more complicated formulas for the cubic and quartic equations (Exercises 1 and 2 of the Sixth Set) have no generalization to the quintic equation.

Having just mentioned the exercises, I hasten to reassure the reader that *the exercises are not essential to the book*. The only proofs that are relegated to the exercises are those that I believe to be too easy, or too much like other proofs already covered, to spend time on in the text. Naturally, the reader who does the exercises will have a far greater understanding of the subject, and will learn many things not contained in the text, but to do all the exercises will surely consume an enormous amount of time. The reader who has just read the text will have covered all the propositions and methods of proof that I consider to be basic to Galois theory.

What preparation do I assume on the part of the reader? Because terminology changes so much from decade to decade and from field to field, I have tried to assume as little terminology as possible. (When I completed my undergraduate degree 25 years ago, I had had courses in advanced calculus, determinants and matrices, differential equations, measure theory, complex variables, etc., but I had never encountered the definition of a group or an abstract vector space.) However, I have assumed a certain degree of mathematical *experience* on the part of the reader, by which I mean experience in computation and mathematical reasoning. The main theorems of Galois theory state, in the last analysis, that certain computations with polynomials produce certain results. In most cases the computations are too long to do, and the *idea* of the computation is what counts, not any particular cases of it. The reader should have enough mathematical experience (and talent) to be able to conceive a general computation and its properties after having done a few simple examples.

The approach of the book is consistently *algebraic* and *constructive*. The fields considered are those obtained from the rational numbers by adjoining a finite number of algebraic and/or transcendental elements. (Fields with characteristic p are mentioned only in passing. Fields obtained by completion processes—the real and complex numbers, algebraic extensions of p-adic fields—are not considered at all.) *The constructive approach implies that theorems mean what they say.* For example, when a theorem says that an equation is solvable, the proof must give a procedure—however impractical—for constructing a splitting field by the adjunction of radicals. I believe that this approach is very much in tune with Galois' conception of the subject.

Liouville, in the "Avertissment" preceding his publication of Galois' works in 1846, writes of the "vivid pleasure" he enjoyed when he realized that Galois' methods were correct and that his theorems could be rigorously proved. I experienced what I imagine was a similar—if lesser—pleasure when I realized that two parts of Galois' memoir, which I at first thought were mistakes, are perfectly correct. These are the places where Galois later wrote "*On jugera*", in the case of the first, and "Something in this proof needs to be completed—I haven't the time" in the case of the second.

The "*On jugera*" passage is the one where Galois proves the crucial lemma stating that any rational function of the roots can be expressed as a rational function of the Galois resolvent. Poisson had called Galois' proof "insufficient" but pointed out that the lemma followed from a theorem of Lagrange. Galois, rather than elucidate his proof, laconically replied, "That remains to be seen" (freely translated). My opinion is in §37.

The famous statement "I haven't the time" occurs in a marginal note Galois made, probably on the night before the duel, with regard to the proof of his Proposition II, which he said needed to be "completed". Although his proof appears wrong at first because he adjoins *one* root r of an equation and then uses *other* roots of the equation, and although Liouville [Gl, p. 492] found it necessary to circumvent Galois' proof entirely, I believe now that the proof given in §44 is very close to what Galois had in mind, and that the marginal note was merely prompted by the fact that he had *changed the statement of the Proposition*, and realized that the proof needed to be amended accordingly. (In fact, the Proposition, as stated, is false. The index of the subgroup need not be 1 or p when p is not prime—it must simply be a divisor of p.) A similar situation occurred with Proposition III, where Galois again changed the statement, making it more general, at the last minute, and had only time enough to say, "One will find the proof."

Finally, I hope it is superfluous to add that, while I have said above that most of what is in this book is already in Galois, the converse is far from true. The book contains a rather complete account of Galois' main memoir, "Mémoire sur les conditions de résolubilité des équations par radicaux" (Appendix 1 contains a translation of this memoir) but it does not make any claim to cover his other works. These contain, I am told, remarkable insights into a number of topics, including the theory of Abelian functions and finite simple groups. I return to my perennial refrain: Read the masters.

Contents

Acknowledgments

My greatest indebtedness is to Mr. James M. Vaughn, Jr., and the Vaughn Foundation Fund. This book is a direct result of their encouragement and support, for which I am deeply grateful. Work on the book was also supported by a Fellowship of the John Simon Guggenheim Memorial Foundation during 1981/82. I am grateful for both the financial support and the honor of a Guggenheim Fellowship. A large number of friends and colleagues have helped me by reading and commenting on early versions of the manuscript. Some major revisions prompted by their criticisms have not been seen by any of them, so it is even truer than usual that they are entitled to credit for many improvements in the book but free from blame for its faults. I would especially like to thank the following for their help: Jay Goldman, Mel Hausner, M.Y. Hirano, Christian Houzel, Susan Landau, Richard Pollack, Walter Purkert, Michael Rosen, Gabriel Stolzenberg, René Taton, William Y. Vélez, B.L. van der Waerden, and an anonymous reader for Springer-Verlag. Finally, my thanks to New York University and the Courant Institute for their overall support and assistance, including a sabbatical year 1980/81, the excellent library, and the expert word-processing of Connie Engle.

Galois

§1   Great mathematicians usually have undramatic lives, or, more precisely, the drama of their lives lies in their mathematics and cannot be appreciated by nonmathematicians. The great exception to this rule is Evariste Galois (1811–1832). Galois' life story—what we know of it—is like a romantic novel. Although he was making important mathematical discoveries when he was still in secondary school, he was denied admission to the Ecole Polytechnique, which was the premier institution of higher learning in mathematics at the time, and the mathematical establishment ignored, mislaid, lost, and failed to understand his treatises. Meanwhile, he was persecuted for his political activities and spent many months in jail as a political prisoner. At the age of 20 he was killed in a duel involving, in some mysterious way, honor and a woman. On the eve of the fatal duel he wrote a letter to a friend outlining his mathematical accomplishments and asking that the friend try to bring his work to the attention of the mathematical world. Against great odds, Galois' few supporters did finally, 14 years after his death, succeed in finding an audience for his work, and portions of his writings were published in 1846 by Joseph Liouville in his *Journal de Mathematiques*. After that, recognition of the great importance of his work came very quickly, and Galois began to be regarded, as he is today, as one of the great creative mathematicians of all time.

§2   The purpose of this book is to convey the mathematical drama of Galois' work, so there will be no more mention of his short, unhappy life,* but

---

* For biographical information see Dupuy [D1], Kiernan [K1], Rothman [R1].

one point needs to be made about its most dramatic feature, namely, the fact that Galois was able, at such an early age and without the benefit of any formal higher education, to make discoveries that would win him lasting fame. Surely many aspiring young mathematicians have been discouraged by Galois' story, saying to themselves something like, "Here I am already x years old, $x - 20$ years older (younger) than Galois was when he *died*, and, although I like math and have always done well at it, I would no more be able to make a great discovery than I would be able to swim the Atlantic." How was Galois able to do it? Was he blessed with some superhuman gift that put him in a class apart? I think not. Of course, talent is essential, and few are as talented as Galois. Still, talent alone is not enough. Galois had to reach the point where he knew enough and had enough techniques at his command to be able to move beyond what had been done before. The secret of how he was able to do this is contained, I believe, in a passage in Dupuy's biography of Galois [D1, p. 206]: "Elementary algebra books never satisfied Galois because he didn't find in them the stamp of the inventors; right from his first year of mathematics he turned to Lagrange."

Lagrange's *Réflexions sur la Résolution Algébrique des Equations* (1771) is the treatise of Lagrange most likely to have inspired the creation of Galois theory. It is an extraordinary work, written in a relaxed, discursive style that was rather common in the eighteenth century, but is virtually unknown in mathematical writing today. It discusses at length the central question of the time in the theory of algebraic equations, namely: What is the essence of the methods by which it is possible to solve equations of degrees 2, 3, and 4? Is it possible to extend these methods to equations of higher degree and, if not, why not? Lagrange gave an insightful answer to the first question, describing the solutions of equations of low degree in terms of a unified technique now known as the technique of the *Lagrange resolvent*.* His answer to the second question, on the other hand, is quite inconclusive. He shows that the technique does not apply in an obvious way to equations of degree 5 or higher, and he discusses some techniques—notably the technique of permuting the roots of an algebraic equation—which are relevant to the applications of Lagrange resolvents to equations of higher degree, but he gives no final answer. In short, it is a paper that gives the reader as much information about the problem as the author can provide and indicates the direction which the author feels further work should take. Viewed in this way, Lagrange's paper seems the perfect source of inspiration for a Galois.

Thus, in order to appreciate Galois' theory, it is natural first to review Lagrange's work. We will go much farther back than that—all the way to ancient Babylon—and then review a few other aspects of the development of algebra before discussing the main features of the work of Lagrange and then moving on to his successors, Gauss and Galois.

---

* A very similar technique was used a few months earlier by Vandermonde (see §15), but this was unknown to Lagrange.

Quadratic Equations 1700 B.C.

§3 Archeological research in the twentieth century has revealed the surprising fact that the peoples of Mesopotamia in the period around* 1700 B.C. had an advanced mathematical culture, including an excellent sexagesimal system of arithmetic and a knowledge of the Pythagorean theorem (a millennium before Pythagoras!). Of particular relevance to the theory of equations and Galois theory is the knowledge in this ancient culture of a method for the solution of quadratic equations.

According to Neugebauer [N1], the technique commonly used in the Babylonian texts to solve quadratic equations can be viewed as a reduction to a normal form, followed by a method for solving the normal form. The normal form was to *find two numbers given their sum and their product.* In modern algebraic notation, this can be stated: Given two numbers p and s, and given that $xy = p, x + y = s$, find x and y. The steps by which the Babylonians solved this problem are as follows:

1. Take half of s.
2. Square the result.
3. From this subtract p.
4. Take the square root of the result.
5. Add this to half of s; this is one of the two numbers and the other is s minus this one.

For example, if the sum is 10 and the product is 21 then the successive steps give 5, 25, 4, 2, 7 and $10 - 7 = 3$. Thus the two numbers are 7 and 3.

That this normal form is indeed a quadratic equation can be seen by multiplying the equation $s = x + y$ by x to find $sx = x^2 + xy = x^2 + p$. In other words, x is a solution of the quadratic equation $x^2 - sx + p = 0$ and, by symmetry, so is y.

Conversely, the solution of any quadratic equation can *in our notation* be viewed as the solution of a problem in normal form. Specifically, the equation $ax^2 + bx + c = 0$ can be rewritten as $x^2 + (c/a) = -(b/a)x$ and the solution of this equation is equivalent to finding two numbers whose sum is $-b/a$ and whose product is c/a. The Babylonians could *not* reduce *all* quadratic equations to a single normal form, however, because their arithmetic did not include negative numbers. To deal with this fact, they had a second normal form, in which the *difference* and the product of two numbers were given. This is a technical problem of considerable historical interest— it was only a few centuries ago that negative numbers became generally accepted so that polynomial equations did not have to be divided into several cases depending on the signs of the coefficients—but is of no importance to the algebra of the problem and will not be considered further here.

---

* The texts cannot be closely dated. Neugebauer places them between 1600 and 1800 B.C.

In modern algebraic notation (also only a few centuries old) the Babylonian solution of the problem in normal form can be written

$$x = \sqrt{\left(\frac{s}{2}\right)^2 - p} + \frac{s}{2}, \qquad y = s - x,$$

or, in a more familiar form,

$$x, y = \frac{s \pm \sqrt{s^2 - 4p}}{2}.$$

Thus it is fair to say that they knew the quadratic formula but that they spelled out the steps of the procedure instead of expressing it as a formula in the way we do.

How did they derive this procedure? Unfortunately, there is no indication in the texts which survive. The point of these texts seems to have been to convey, by means of several worked examples, the technique of solution. It is entirely possible that the technique was discovered by an ancient genius and that his successors merely adopted it because it produced correct answers. On the other hand, it may be that some derivation was well understood by many people at the time, but was transmitted orally or does not happen to be among the texts that have been found.

Cubic and Quartic Equations A.D. 1500

§4   There was some progress in algebra in the 3000 years between the Old Babylonian period and the Italian Renaissance, but it was not great. The late Greek writer Diophantus (circa A.D. 250) introduced some abbreviated algebraic notation, the Hindus used negative numbers on occasion, and the Arabs constructed the solutions of cubic equations as points of intersection of conic sections. When the Renaissance came, however, the advances in algebra were enormous, and they opened the way to great progress in all branches of mathematics.

In mathematics, the Renaissance was not a rebirth at all, but a period of first vigorous growth. In ancient times, Europe had been a mathematical backwater, and even the Romans were barbarians when it came to mathematics. During the Middle Ages, Europe had learned about algebra (al-jabr) from the Arabs and had begun to improve it by devising new symbols and notations. Then, in the sixteenth century, an enormous advance was made— the algebraic solution of cubic equations was discovered, and soon thereafter the solution of quartic equations.

The history of the discovery of these solutions and their exact description in terms of the still quite clumsy notation of the period will not be necessary in what follows. Instead, we will give just a brief account, in modern notation, of the solutions themselves. (For more details see Kline [K2], pp. 263–270 and 282–284.)

§5   Suppose the cubic equation to be solved has the form $x^3 + px + q = 0$. (An arbitrary cubic equation can be put in this form by dividing by the coefficient of x^3 and then taking a change of variable $x' = x - c$ with c equal to the coefficient of x^2 divided by 3.) Introduce two new variables a and b and set $x = a - b$. The desired equation is then $a^3 - 3a^2b + 3ab^2 - b^3 + pa - pb + q = 0$, that is, $a^3 - b^3 + (a - b)(-3ab + p) + q = 0$. If it is stipulated that $3ab = p$, then this equation takes the form $a^3 - b^3 + q = 0$. If a solution (a, b) of these two equations $3ab = p$ and $a^3 - b^3 + q = 0$ in two unknowns* can be found, then, as is easily checked, the quantity $x = a - b$ is a solution of the original equation $x^3 + px + q = 0$. Multiplication by 3^3a^3 makes it possible to eliminate b from $a^3 - b^3 + q = 0$ to find $27a^6 - (3ab)^3 + 27a^3q = 0$, that is, $27a^6 + 27qa^3 - p^3 = 0$. This is a quadratic equation for a^3. Let a be the cube root of a solution of this quadratic equation and let $b = p/3a$. Then $3ab = p$ and $a^3 - b^3 + q = 0$, which implies that $x = a - b$ is a solution of the given equation.

§6 For the solution of the quartic, assume that the equation has the form $x^4 + px^2 + qx + r = 0$. (Again, an arbitrary quartic equation can easily be put in this form.) Let this be put in the form $x^4 = -px^2 - qx - r$. Then, if a is a new variable, $(x^2 + a)^2 = x^4 + 2ax^2 + a^2 = (-p + 2a)x^2 - qx + (-r + a^2)$. In order to take a square root on the right side, this quadratic function of x should have a single root—i.e. should be of the form $A(x + B)^2$—and by the quadratic formula this occurs if and only if $q^2 - 4(-p + 2a)(-r + a^2) = 0$. This is a cubic equation for a, which can (by the above method) be solved for a. When a is a root of this equation, the right side of the above expression of $(x^2 + a)^2$ has the form $A(x + B)^2$ where A is the coefficient of x^2 and B is the coefficient of x divided by $2A$, that is,

$$(x^2 + a)^2 = (-p + 2a)\left(x - \frac{q}{2(-p + 2a)}\right)^2,$$

or, more simply,

$$x^2 + a = \pm\sqrt{-p + 2a}\left(x - \frac{q}{2(-p + 2a)}\right),$$

which gives x as the solution of a quadratic equation.

§7 Of course the successful solution of the cubic and quartic equations led to attempts to solve the quintic equation. It was not until almost 300 years later, in the 1820's, that it was shown, first by Abel, then by Galois, that it is *impossible* to solve the quintic equation in the same manner that the cubic and the quartic were solved, specifically, by using no operations other than addition, subtraction, multiplication, division, and the extraction of roots.

During these 300 years the fruitful developments in algebra were in other

---

* There is no sharp distinction made here among the terms "variable", "unknown", and "indeterminate". For the most part, "variable" is used in this book. If a variable occurs in an equation that is to be solved, it may be called an unknown. If it is to remain variable and is being used primarily as a placeholder in a computation, it may be called an indeterminate.

directions. One of the most important was a theorem discovered by Isaac Newton, which is the subject of the next section.

Newton and Symmetric Functions

§8   Isaac Newton (1643–1727) is most famous for his discovery of the universal law of gravitation and for his use of that law to give an exact mathematical description of planetary motion. Consequently, he is identified in most people's minds with mathematical physics and applied mathematics. Even people who have some acquaintance with the history of mathematics and who realize that Newton, with Leibniz, is regarded as the father of differential and integral calculus, tend to think of Newton's mathematics as being closely related to his physics, and his calculus as being primarily a tool in his study of motion. Nevertheless, Newton's contributions to pure mathematics alone are sufficient to place him among the greatest geniuses in the history of mathematics. This section is devoted to a theorem of pure algebra which is of crucial importance to the later development of the subject and which appears to be Newton's creation.

A portion of this theorem was published in Newton's *Arithmetica Universalis* in 1707, after Newton was world famous and had ceased active scientific work. It is cited by Gauss [G2, Art. 338] and Weber [W3, vol. I, Sec. 46], among others, and is generally known as Newton's theorem. Of course the *Arithmetica Universalis* was known to have been written long before 1707, but it is only with the recent work of Derek T. Whiteside in analyzing, annotating, and publishing Newton's notebooks and papers that it has been possible to date many of Newton's discoveries and, in the case of the theorem under discussion, to know that he was aware at a very early date of the *full* theorem, not just the portion given in the *Arithmetica Universalis*.

§9   Whiteside found in papers dating to 1665–1666, in the very earliest phase of Newton's career, the following formulas: Let r, s, t be the three roots of a cubic equation $x^3 + bx^2 + cx + d = 0$, and let an expression like "every $r^i s^j$" denote the sum of all distinct expressions of the form $r^i s^j$ where r and s are roots of the given cubic, i.e. "every $r^2 s$" $= r^2 s + s^2 t + t^2 r + r^2 t + t^2 s + s^2 r$, "every r^2" $= r^2 + s^2 + t^2$, "every $r^2 s^2 t^2$" $= r^2 s^2 t^2$, etc. Then Newton's formulas* are

$$(\text{every } r) = -b \tag{1}$$

$$(\text{every } r^2) = b^2 - 2c \tag{2}$$

$$(\text{every } r^3) = -b^3 + 3bc - 3d \tag{3}$$

$$(\text{every } rs) = c \tag{4}$$

$$(\text{every } r^2 s) = -bc + 3d \tag{5}$$

$$(\text{every } r^3 s) = b^2 c - 2c^2 - bd \tag{6}$$

---

* [N3, p. 517]. Newton took $-r, -s, -t$ to be the roots of the equation, which simply changes the signs of the formulas with odd degree.

$$(\text{every } r^2s^2) = c^2 - 2bd \tag{7}$$

$$(\text{every } r^3s^2) = -bc^2 + 2b^2d + cd \tag{8}$$

$$(\text{every } r^3s^3) = c^3 - 3bcd + 3d^2 \tag{9}$$

$$(\text{every } rst) = -d \tag{10}$$

$$(\text{every } r^2st) = bd \tag{11}$$

$$(\text{every } r^3st) = -b^2d + 2cd \tag{12}$$

$$(\text{every } r^2s^2t) = -cd \tag{13}$$

$$(\text{every } r^3s^2t) = bcd - 3d^2 \tag{14}$$

$$(\text{every } r^3s^3t) = -c^2d + 2bd^2 \tag{15}$$

$$(\text{every } r^2s^2t^2) = d^2 \tag{16}$$

$$(\text{every } r^3s^2t^2) = -bd^2 \tag{17}$$

$$(\text{every } r^3s^3t^2) = cd^2 \tag{18}$$

$$(\text{every } r^3s^3t^3) = -d^3 \tag{19}$$

He did not record in his notes the method by which he derived these formulas, and we can only guess what lay behind them. However, it seems likely that the choice to stop with third powers of the roots was arbitrary* and that he could have given analogous formulas for higher powers. Moreover, the decision to deal with the three roots of a cubic, rather than the four roots of a quartic or the five roots of a quintic, was also probably arbitrary. In fact, a few pages later in Whiteside's book, a passage from Newton's notebook is reproduced in which he gives the analogs of formulas (1)–(3) for an equation of degree 8, namely, the formulas†

$(\text{every } r) = -p,$

$(\text{every } r^2) = p^2 - 2q,$

$(\text{every } r^3) = -p^3 + 3pq - 3r,$

$(\text{every } r^4) = p^4 - 4p^2q + 4pr - 4s + 2q^2,$

$(\text{every } r^5) = -p^5 + 5p^3q - 5p^2r + 5ps - 5t - 5pq^2 + 5qr,$

$(\text{every } r^6) = p^6 - 6p^4q + 6p^3r - 6p^2s + 6pt - 6v + 9p^2q^2$
$\qquad\qquad\quad - 12pqr + 6qs - 2q^3,$

$(\text{every } r^7) = -p^7 + 7p^5q - 7p^4r + 7p^3s - 7p^2t + 7pv - 7y,$

$(\text{every } r^8) = p^8 - 8p^6q + 8p^5r - 8p^4s + 8p^3t - 8p^2v + 8py - 8z,$

* Newton in fact had a specific goal in mind in the passage in question, namely, the derivation of the explicit formula for the resultant of two cubics (see Exercise 8). For this goal he needed the given formulas and only these.

† The first four of these formulas were published by Albert Girard in 1629. In Whiteside's opinion, Newton was not aware of Girard's work. The formulas at the bottom of this page are taken from Whiteside's edition of Newton's papers (vol. 1, p. 519). They are incorrect in the cases 6, 7, and 8. The formula for "every r^6" is missing a single term $3r^2$. (Note that there is no u—in Newton's alphabet, t is followed by v.) The correct formulas for "every r^7" and "every r^8" are much longer than the formulas given here. I do not know whether the errors are Newton's or Whiteside's.

where r runs over the eight roots of the 8th degree equation $x^8 + px^7 + qx^6 + rx^5 + sx^4 + tx^3 + vx^2 + yx + z = 0$.

In other words, it appears likely that Newton was aware that there are analogous formulas for all degrees, that is, that *any symmetric polynomial in the roots of an equation can be expressed in terms of the coefficients of that equation*. This theorem is the foundation stone of Galois theory, so it is important to have a careful statement and proof of it before proceeding. (It must be admitted, however, that neither a careful statement nor a proof of it seems to have been published before the nineteenth century. Everyone seemed familiar with it and used it without inhibition.)

The Fundamental Theorem on Symmetric Polynomials

§10   The first step in giving a careful statement of the theorem is to remove the reference to roots of an nth degree equation, because these roots may be irrational or complex and they are really extraneous to the theorem. (Newton explicitly states in his formulas that the roots may be "false", i.e. negative, or "imaginary".) The particular formulas (1), (4) and (10) in Newton's list, that is,

$$
\begin{aligned}
r + s + t &= -b, \\
rs + st + tr &= c, \\
rst &= -d,
\end{aligned}
\tag{20}
$$

are especially important and were probably rather widely known in Newton's time. (Whiteside [N3, p. 518, note 12] observes that the general case of these formulas was published by Albert Girard in 1629 but says that "we may assume" that Newton's version of it, which he published in the *Arithmetica Universalis*, was his "independent discovery".) These formulas follow immediately from the identity

$$
x^3 + bx^2 + cx + d = (x - r)(x - s)(x - t),
$$

when the right side is multiplied out and coefficients of like powers of x are equated. The same procedure applied to

$$
x^n + b_1 x^{n-1} + b_2 x^{n-2} + \cdots + b_n = (x - r_1)(x - r_2) \cdots (x - r_n)
$$

shows that, in analogy to (20), the sum of all* $\binom{n}{k}$ products of k of the r_i is equal to $(-1)^k b_k$. That is,

$$
\begin{aligned}
r_1 + r_2 + \cdots + r_n &= -b_1, \\
r_1 r_2 + r_1 r_3 + \cdots + r_{n-1} r_n &= b_2, \\
r_1 r_2 r_3 + r_1 r_2 r_4 + \cdots + r_{n-2} r_{n-1} r_n &= -b_3, \\
&\vdots \\
r_1 r_2 \cdots r_n &= (-1)^n b_n.
\end{aligned}
$$

* Here $\binom{n}{k}$ denotes the binomial coefficient $\dfrac{n!}{k!\,(n-k)!}$.

Let σ_k denote the polynomial on the left side of the kth equation, so that the equation reads $\sigma_k = (-1)^k b_k$. Then σ_k is called *the kth elementary symmetric polynomial* in $r_1, r_2, \ldots, r_n$. Here $r_1, r_2, \ldots, r_n$ are to be regarded as variables or indeterminates, rather than roots of an equation, and $\sigma_1, \sigma_2, \ldots, \sigma_n$ are to be regarded as polynomials in these indeterminates. A polynomial in $r_1, r_2, \ldots, r_n$ is simply a formal sum of terms of the form $A r_1^{\mu_1} r_2^{\mu_2} \cdots r_n^{\mu_n}$ where A is a number (say a rational number for now) and the μ's are nonnegative integers. A polynomial in $r_1, r_2, \ldots, r_n$ is said to be *symmetric* if it has the property that interchanging any two of the r's leaves the polynomial unchanged—provided, of course, that two polynomials are regarded as being equal when applications of the commutative laws of addition and multiplication can change one into the other. (For example, $rs + st + tr$ is unchanged by an interchange of r and s because $rs + st + tr = sr + rt + ts$.) Clearly the elementary symmetric polynomials $\sigma_1, \sigma_2, \ldots, \sigma_n$ in n variables are symmetric in this definition.

In Newton's formulas (1)–(19) the left sides are by definition symmetric in r, s, and t, and the right sides are polynomials in b, c, and d or, what is the same in view of (20), polynomials in $\sigma_1 = r + s + t$, $\sigma_2 = rs + st + tr$, and $\sigma_3 = rst$. Thus these formulas are all instances of the following theorem:

Fundamental Theorem on Symmetric Polynomials. *Every symmetric polynomial in $r_1, r_2, \ldots, r_n$ can be expressed as a polynomial in the elementary symmetric polynomials $\sigma_1, \sigma_2, \ldots, \sigma_n$. Moreover, a symmetric polynomial with integer coefficients can be expressed as a polynomial in $\sigma_1, \sigma_2, \ldots, \sigma_n$ with integer coefficients.*

When the theorem is stated in this way, it deals with algebraic identities, not with equations and roots. For example, Newton's formula (3) is the identity

$$r^3 + s^3 + t^3 = \sigma_1^3 - 3\sigma_1\sigma_2 + 3\sigma_3$$
$$= (r + s + t)^3 - 3(r + s + t)(rs + st + tr) + 3rst.$$

However, if $x^n + b_1 x^{n-1} + b_2 x^{n-2} + \cdots + b_n = 0$ is an nth degree equation with n roots (in some sense) the formula $\sigma_k = (-1)^k b_k$ makes it possible to evaluate the elementary symmetric functions of the roots *immediately*, without ever finding the roots themselves; then the theorem shows that it is possible to evaluate *any* symmetric function of the roots without finding the roots themselves. Moreover—a fact that will be very important later on—if the b's are rational numbers, then any symmetric polynomial in the roots assumes a rational value.

§11 PROOF. The proof of the fundamental theorem on symmetric polynomials which follows is a simple computational procedure for finding, given a symmetric polynomial, a representation in terms of elementary symmetric polynomials. To say that the procedure is simple is not to say that it is practical, and in fact the procedure normally leads to an impossibly long computation. This is to be expected when a computational procedure is used to prove a nontrivial existence theorem; a simple and sure, but probably very slow, method is best for proving existence, but, once existence has been established, there are frequently ways to get to the desired answer much more quickly in particular cases.

The proof will be by induction on the number of variables. If there is only one variable, then there is only one elementary symmetric polynomial $\sigma_1 = r_1$ and the theorem is trivially true. Assume, therefore, that the theorem has been proved for symmetric polynomials in $n - 1$ variables and let $G(r_1, r_2, \ldots, r_n)$ be a symmetric polynomial in n variables. Let the terms of G be gathered according to the power of r_n that they contain, to give

$$G(r_1, r_2, \ldots, r_n) = G_0 + G_1 r_n + G_2 r_n^2 + \cdots + G_v r_n^v,$$

where G_i $(i = 0, 1, \ldots, v)$ is a polynomial in the $n - 1$ variables $r_1, r_2, \ldots, r_{n-1}$ and v is the highest power of r_n that occurs in G. Now since G is unchanged when any two of the variables $r_1, r_2, \ldots, r_{n-1}$ are interchanged, so is each G_i. Therefore, by the induction hypothesis, each G_i can be written as a polynomial in the elementary symmetric functions in $r_1, r_2, \ldots, r_{n-1}$. Let $\tau_1, \tau_2, \ldots, \tau_{n-1}$ denote these elementary symmetric functions, that is,

$$\tau_1 = r_1 + r_2 + \cdots + r_{n-1},$$

$$\tau_2 = r_1 r_2 + r_1 r_3 + \cdots + r_{n-2} r_{n-1},$$

$$\vdots$$

$$\tau_{n-1} = r_1 r_2 \cdots r_{n-1}.$$

Then each G_i can be expressed as a polynomial in the τ's, say

$$G_i(r_1, r_2, \ldots, r_{n-1}) = g_i(\tau_1, \tau_2, \ldots, \tau_{n-1})$$

and, in addition, if G has integer coefficients then so do all the polynomials g_i.

As before, let $\sigma_1, \sigma_2, \ldots, \sigma_n$ be the elementary symmetric functions in n variables. Then clearly the σ's and the τ's satisfy the relations

$$\sigma_1 = \tau_1 + r_n,$$

$$\sigma_2 = \tau_2 + r_n \tau_1,$$

$$\sigma_3 = \tau_3 + r_n \tau_2,$$

$$\sigma_4 = \tau_4 + r_n \tau_3,$$

$$\vdots$$

$$\sigma_n = 0 + r_n \tau_{n-1},$$

and these imply

$$\tau_1 = \sigma_1 - r_n,$$

$$\tau_2 = \sigma_2 - r_n\tau_1 = \sigma_2 - r_n\sigma_1 + r_n^2,$$

$$\tau_3 = \sigma_3 - r_n\tau_2 = \sigma_3 - r_n\sigma_2 + r_n^2\sigma_1 - r_n^3,$$

$$\vdots$$

$$\tau_{n-1} = \sigma_{n-1} - r_n\tau_{n-2} = \sigma_{n-1} - r_n\sigma_{n-2} + \cdots + (-1)^{n-1}r_n^{n-1},$$

$$0 = \sigma_n - r_n\tau_{n-1} = \sigma_n - r_n\sigma_{n-1} + \cdots + (-1)^n r_n^n.$$

Let these expressions for $\tau_1, \tau_2, \ldots, \tau_{n-1}$ in terms of σ's and r_n be substituted into the equation

$$G = g_0(\tau) + g_1(\tau)r_n + g_2(\tau)r_n^2 + \cdots + g_\nu(\tau)r_n^\nu.$$

The result is a polynomial in $\sigma_1, \sigma_2, \ldots, \sigma_{n-1}$ and r_n. Let the terms be gathered according to the power of r_n that they contain to give an expression of the form

$$G(r_1, r_2, \ldots, r_n) = f_0(\sigma) + f_1(\sigma)r_n + f_2(\sigma)r_n^2 + \cdots + f_n(\sigma)r_n^\mu$$

where each $f_i(\sigma)$ is a polynomial in $\sigma_1, \sigma_2, \ldots, \sigma_{n-1}$ which has integer coefficients if G does. If $\mu \geq n$ then the degree of this polynomial with respect to r_n can be reduced using the relation

$$r_n^n = r_n^{n-1}\sigma_1 - r_n^{n-2}\sigma_2 + r_n^{n-3}\sigma_3 - \cdots + (-1)^{n-1}\sigma_n$$

found above. Therefore, by repeated use of this identity, the degree in r_n can be reduced to $n - 1$ and an identity of the form

$$G(r_1, r_2, \ldots, r_n) = f_0(\sigma) + f_1(\sigma)r_n + \cdots + f_{n-1}(\sigma)r_n^{n-1} \qquad (1)$$

can be derived. Here each $f_i(\sigma)$ is a polynomial in $\sigma_1, \sigma_2, \ldots, \sigma_n$ which has integer coefficients if G has integer coefficients. The proof will be completed by showing that, in a relation of this form, if G is symmetric, then

$$f_1, f_2, \ldots, f_{n-1}$$

are all necessarily 0 so that $G(r_1, r_2, \ldots, r_n) = f_0(\sigma_1, \sigma_2, \ldots, \sigma_n)$ as was to be shown.

Both sides of (1) are polynomials in $r_1, r_2, \ldots, r_n$. Therefore it is meaningful to interchange r_1 and r_n in each side. This leaves G unchanged, by assumption, and leaves $f_i(\sigma)$ unchanged because the σ's are symmetric. Therefore it changes (1) into the same equation with r_1 in place of r_n. In the same way,

r_n can be interchanged with any one of $r_1, r_2, \ldots, r_{n-1}$. This gives an $n \times n$ system of linear equations

$$f_0(\sigma) + f_1(\sigma)r_1 + f_2(\sigma)r_1^2 + \cdots + f_{n-1}(\sigma)r_1^{n-1} = G(r_1, r_2, \ldots, r_n),$$
$$f_0(\sigma) + f_1(\sigma)r_2 + f_2(\sigma)r_2^2 + \cdots + f_{n-1}(\sigma)r_2^{n-1} = G(r_1, r_2, \ldots, r_n),$$
$$\vdots$$
$$f_0(\sigma) + f_1(\sigma)r_n + f_2(\sigma)r_n^2 + \cdots + f_{n-1}(\sigma)r_n^{n-1} = G(r_1, r_2, \ldots, r_n),$$

in which the matrix of coefficients is $[r_i^{j-1}]$, that is, has r_i^{j-1} in the ith row and jth column. The determinant of this matrix, regarded as a polynomial in $r_1, r_2, \ldots, r_n$, is obviously nonzero because no two of its $n!$ terms contain the r's to the same powers.* Thus the fact that multiplication of the column vector $(f_0(\sigma), f_1(\sigma), \ldots, f_{n-1}(\sigma))$ by this matrix gives the same result as multiplication of the column vector $(G, 0, 0, \ldots, 0)$ by the same matrix implies that these two column vectors are identical, as was claimed above (see Exercise 30).   □

A slightly different way to reach the desired conclusion $G = f_0$ is to regard the above n equations as saying that the polynomial $F(X) = f_{n-1}X^{n-1} + f_{n-2}X^{n-2} + \cdots + f_1 X + f_0 - G$ (with coefficients that are polynomials in $r_1, r_2, \ldots, r_n$) has at least the n roots $X = r_1, r_2, \ldots, r_n$. Therefore (Exercise 29) it is divisible by $(X - r_1)(X - r_2)\cdots(X - r_n)$. This would imply $\deg F \geq n$ if $F(X)$ were not the zero polynomial, so $F(X) = 0$, which is to say $f_0 = G, f_1 = 0, f_2 = 0, \ldots, f_{n-1} = 0$.

Particular Symmetric Polynomials

§12   It was mentioned above that there is a theorem connected with the fundamental theorem on symmetric polynomials that is traditionally known as "Newton's theorem." Specifically, this theorem from the *Arithmetica Universalis*† is the following. For a given value of n, let $\sigma_1, \sigma_2, \ldots, \sigma_n$ be the elementary symmetric functions in n variables $r_1, r_2, \ldots, r_n$, and let s_k, for $k = 1, 2, 3, \ldots$ be the symmetric polynomial

$$s_k = r_1^k + r_2^k + \cdots + r_n^k.$$

By the fundamental theorem, each s_k can be expressed in terms of the σ's. In many situations it is useful to be able to find these expressions explicitly. (When $n = 8$ they are given in §9 above.) "Newton's theorem" is the relation

$$s_k - s_{k-1}\sigma_1 + s_{k-2}\sigma_2 - \cdots + (-1)^{k-1}s_1\sigma_{k-1} + (-1)^k k\sigma_k = 0$$
$$(k = 1, 2, 3, \ldots),$$

---

* This is the "Vandermonde determinant." See note to §15.

† [N2, p. 203 in the English translation.]

where, as is natural, $\sigma_j = 0$ for $j > n$. Thus

$$s_1 - \sigma_1 = 0,$$

$$s_2 - s_1\sigma_1 + 2\sigma_2 = 0,$$

$$s_3 - s_2\sigma_1 + s_1\sigma_2 - 3\sigma_3 = 0,$$

etc.

These relations make it easy to compute the s_k when the σ_k are known, and vice versa. In particular, given any polynomial, one can easily find the sum of the kth powers of its roots without finding the roots. (See Exercises 15 and 16.)

§13 A symmetric polynomial of particular importance is the *discriminant* $\prod (r_i - r_j)^2$ where i and j run over all pairs of distinct integers $1 \le i < j \le n$. In the case $n = 2$ the discriminant is $(r_1 - r_2)^2 = r_1^2 - 2r_1r_2 + r_2^2 = \sigma_1^2 - 4\sigma_2$, a formula which lies at the base of the solution of the quadratic equation. In particular, the discriminant of the roots of a quadratic equation $x^2 + bx + c = 0$ is $(-b)^2 - 4c$, which shows that the two roots are equal if and only if $b^2 - 4c = 0$. For more variables, the discriminant is increasingly difficult to express explicitly in terms of the σ's. For example, for $n = 3$ it is $\sigma_1^2\sigma_2^2 - 4\sigma_1^3\sigma_3 + 18\sigma_1\sigma_2\sigma_3 - 4\sigma_2^3 - 27\sigma_3^2$ (see Exercise 25). Thus, in particular, an equation of the form $x^3 + px + q = 0$ has multiple roots (its discriminant is 0) if and only if $-4p^3 - 27q^2 = 0$.

First Exercise Set

1. Show that if x and y are any two (real) numbers, and if $s = x + y$, $p = xy$, then $s^2 - 4p \ge 0$. Thus if $s^2 - 4p < 0$ the problem $s = x + y$, $p = xy$ has no (real) solution.

2. Show that the formula

$$x, y = \frac{s \pm \sqrt{s^2 - 4p}}{2}$$

of §3 solves the problem *algebraically* in the sense that it gives a solution whenever $\sqrt{s^2 - 4p}$ is something with the property that its square is $s^2 - 4p$. Thus, in particular, it is valid for $s^2 - 4p < 0$ when $\sqrt{s^2 - 4p}$ is regarded as a complex number.

3. The solution of the cubic in §5 should also be regarded as *algebraic*. It is not well suited to the numerical solution of equations. Show that for the solution of $x^3 + 6x = 20$ it gives $x = \sqrt[3]{10 + \sqrt{108}} - \sqrt[3]{-10 + \sqrt{108}}$. To derive the solution $x = 2$, put $\sqrt[3]{\pm 10 + \sqrt{108}}$ in simpler form by solving $(c + d\sqrt{3})^3 = \pm 10 + 6\sqrt{3}$ for c and d. (This example is from Cardan's *Ars Magna*.) Show that the other two solutions of $x^3 + 6x = 20$ are not real.

4. Solve $x^3 = 3x + 2$ by the method of §5. Use the solution that is obtained to find a linear factor of $x^3 - 3x - 2$ and thereby to find two other real roots of the equation.

Show that these roots are given directly by the method of §5 if one uses the *complex* roots of $\omega^3 = 1$, that is, $\omega = \frac{1}{2}[-1 \pm i\sqrt{3}]$.

5. Find the complex roots of the equation of Exercise 3 by: (a) solving the quadratic equation they satisfy; and (b) using complex values of a in the solution of the cubic.

6. Show that the solution of an nth degree equation of the form $x^n + ax^{n-1} + bx^{n-2} + \cdots + dx + e = 0$ can be reduced to the solution of another such equation in which $a = 0$ by the substitution $x = y + C$.

7. Show that the method of §6 solves $x^4 + px^2 + qx + r = 0$ algebraically in the sense that it writes the left side of this equation as a product of two factors of degree 2 (which can then be written as products of linear factors using the quadratic formula).

8. Newton's objective in the passage where he gives the first set of formulas in §9 is to *determine, given two cubic polynomials, whether they have a common root*. Let $f(x) = x^3 + bx^2 + cx + d$ have roots r, s, t, as in §9, and let $g(x) = x^3 + Bx^2 + Cx + D$. Then $g(x)$ and $f(x)$ have a root in common if and only if $g(r)g(s)g(t) = 0$. Since $g(r)g(s)g(t)$ is a symmetric polynomial in r, s, t whose coefficients are polynomials in B, C, D with integer coefficients, this condition can be expressed in the form $P(b, c, d, B, C, D) = 0$ where P is a polynomial with integer coefficients. Find P. [An easy but slightly long computation. P has 34 terms.]

9. The polynomial P of the preceding exercise is called the *resultant* of $x^3 + bx^2 + cx + d$ and $x^3 + Bx^2 + Cx + D$. This gives the resultant of $x^3 + (b/a)x^2 + (c/a)x + (d/a)$ and $x^3 + (B/A)x^2 + (C/A)x + (D/A)$ which, because these polynomials have the same roots, it is natural to call the resultant of $ax^3 + bx^2 + cx + d$ and $Ax^3 + Bx^2 + Cx + D$. Find this resultant. [Easy from Exercise 8.] This, in essense, is the form in which Newton gives the answer. Prove that the resultant is 0 if and only if the two polynomials have a root in common *or* $A = a = 0$, i.e. *neither* polynomial is in fact cubic.

10. Use the resultant to reduce the solution of two equations in two variables $f(x, y) = 0$, $g(x, y) = 0$ to the solution of polynomials in one variable in the case where f and g both have degree 3 at most in x. In all likelihood, this is the application that Newton had in mind and it is the reason he did not assume $a = A = 1$. (Rather rough arguments are all one can expect here, since the notion of a *solution* of an equation has yet to be spelled out.)

11. Given a polynomial in n variables $f(x_1, x_2, \ldots, x_n)$, let its terms be ordered as follows. Arrange terms first according to the power of x_1 they contain: $f = A_h x_1^h + A_{h-1} x_1^{h-1} + \cdots + A_1 x_1 + A_0$ where the A's are polynomials in $x_2, x_3, \ldots, x_n$. Next arrange the terms of A_i according to the powers of x_2 they contain: $A_i = B_{ik} x_2^k + \cdots + B_{i1} x_2 + B_{i0}$, then arrange the terms of B_{ij} according to the powers of x_3 they contain, etc. Otherwise stated, put the terms of f in *lexicographic order*, where one term precedes another in lexicographic order if it contains x_1 to a higher power, or, if they contain x_1 to the same power, it contains x_2 to a higher power, or, if they contain x_1 and x_2 to the same power, it contains x_3 to a higher power, etc. Show that if f and g are two polynomials in $x_1, x_2, \ldots, x_n$ in lexicographic order then the leading term of fg is equal to the leading term of f times the leading term of g.

12. Give a proof of the fundamental theorem on symmetric polynomials along the following lines. Let $G(x_1, x_2, \ldots, x_n)$ be a symmetric polynomial, and let $A x_1^{m_1} x_2^{m_2} \cdots x_n^{m_n}$

be its leading term (in lexicographic order). Then $m_1 \geq m_2 \geq \cdots \geq m_n$ and

$$A\sigma_1^{m_1-m_2}\sigma_2^{m_2-m_3}\cdots\sigma_{n-1}^{m_{n-1}-m_n}\sigma_n^{m_n} = f$$

is a symmetric polynomial with the same leading term. Similarly, let g be the monomial in $\sigma_1, \sigma_2, \ldots, \sigma_n$ with the same leading term as $G - f$, h the monomial with the same leading term as $G - f - g$, and so forth. Show that eventually one of these differences must be 0, and therefore that G can be written as a sum of monomials in $\sigma_1, \sigma_2, \ldots, \sigma_n$ with coefficients that are integers if the coefficients of G are integers.

13. Use the method of the preceding exercise to derive formulas (1)–(19) of §9.

14. Give an alternate proof of the identity $r_n^n - \sigma_1 r_n^{n-1} + \sigma_2 r_n^{n-2} - \cdots \pm \sigma_n = 0$ of §11.

15. Prove *Newton's Theorem* (§12).

16. Use Newton's Theorem to derive the formulas for "every r^i" ($i = 1, 2, \ldots, 8$) in §9.

17. Find an equation of degree 2 whose roots are the *cubes* of the roots of $x^2 + ax + b$. Apply this to the equation $x^2 + x + 1$ to conclude that its roots are complex cube roots of unity.

18. Find all quadratic equations $x^2 + ax + b = 0$ with the property that the squares of the roots coincide with the roots.

19. For a given set $x_1, x_2, \ldots, x_n$ of n variables let $(k_1, k_2, \ldots, k_h)$ denote the polynomial which Newton would call "every $x_1^{k_1} x_2^{k_2} \cdots x_h^{k_h}$," that is, the sum of all *distinct* monomials that can be obtained from $x_1^{k_1} x_2^{k_2} \cdots x_h^{k_h}$ by permutations of $x_1, x_2, \ldots, x_n$. (If $h > n$ let $(k_1, k_2, \ldots, k_h)$ be the 0 polynomial.) Prove the following rule for computing $(k_1, k_2, \ldots, k_h)(m)$. Let A be the set of all functions $(k_1', k_2', \ldots, k_{h+1}')$ such that (1) $k_1' \geq k_2' \geq \cdots \geq k_{h+1}'$ and (2) the $(h+1)$-tuple of integers $(k_1', k_2', \ldots, k_{h+1}')$ is the componentwise sum of two $(h+1)$-tuples, one of which is a permutation of $(k_1, k_2, \ldots, k_h, 0)$ and the other of which is a permutation of $(m, 0, 0, \ldots, 0)$. Then $(k_1, k_2, \ldots, k_h)(m) = \sum_{f \in A} c_f f$ where c_f is the number of *distinct* ways that the k-tuple $f = (k_1', k_2', \ldots, k_{h+1}')$ can be written as a sum of two $(h+1)$-tuples in the prescribed manner, or, what is the same, the number of terms in the sequence $f = (k_1', k_2', \ldots, k_{h+1}')$ that are equal to the value k_i' which contains m. For example, $(1, 1)(1)$ gives $A = (2, 1, 0)$, $(1, 1, 1)$; the first arises only as $(1, 1, 0) + (1, 0, 0)$, but the second arises three ways— $(1, 1, 0) + (0, 0, 1)$, $(1, 0, 1) + (0, 1, 0)$, and $(0, 1, 1) + (1, 0, 0)$. Therefore $(1, 1)(1) = (2, 1) + 3(1, 1, 1)$.

20. Using the preceding exercise, show that every symmetric polynomial with integer coefficients can be written as a polynomial in $s_i = x_1^i + x_2^i + \cdots + x_n^i$ ($i = 1, 2, 3, \ldots$) with *rational* coefficients.

21. Prove (6) of §9 by expressing (3, 1) in terms of $s_1 = r + s + t$, $s_2 = r^2 + s^2 + t^2$, $s_3 = r^3 + s^3 + t^3, \ldots$ as in the preceding exercise, and expressing $s_1, s_2, s_3, \ldots$ in terms of σ's.

22. Show that the representation in the fundamental theorem on symmetric polynomials is *unique*. In other words, show that if $F(y_1, y_2, \ldots, y_n)$ is a polynomial in n variables with the property that substitution of the symmetric functions $\sigma_1, \sigma_2, \ldots, \sigma_n$ in n variables $x_1, x_2, \ldots, x_n$ in F gives the 0 polynomial $F(\sigma_1, \sigma_2, \ldots, \sigma_n) = 0$, then F is the 0 polynomial.

23. Prove the formula

$$\begin{vmatrix} 1 & 1 & 1 & 1 \\ a & b & c & d \\ a^2 & b^2 & c^2 & d^2 \\ a^3 & b^3 & c^3 & d^3 \end{vmatrix} = (d - c)(d - b)(d - a)(c - b)(c - a)(b - a)$$

for the 4 × 4 Vandermonde determinant and its generalization to the $n \times n$ case.

24. Multiply a Vandermonde determinant by its transpose to prove the formula

$$\begin{vmatrix} s_0 & s_1 & s_2 & s_3 \\ s_1 & s_2 & s_3 & s_4 \\ s_2 & s_3 & s_4 & s_5 \\ s_3 & s_4 & s_5 & s_6 \end{vmatrix} = D$$

for the discriminant D of a 4th degree equation (see §13).

25. Use the analog of the formula of Exercise 24 to derive the formula for the discriminant of a cubic given in §13.

26. Show that if $f(x) = (x - r_1)(x - r_2)(x - r_3)$ then the discriminant is the negative of the resultant of f and its derivative f'. Use the formula of Exercise 8 to derive the formula of §13.

27. Derive the formulas for $s_1, s_2, \ldots, s_8$ in §9 by taking the log of $(1 + r_1)(1 + r_2) \cdots (1 + r_n) = 1 + \sigma_1 + \sigma_2 + \cdots + \sigma_n$ using the series $\log(1 + x) = x - \frac{1}{2}x^2 + \frac{1}{3}x^3 - \frac{1}{4}x^4 + \cdots$ and equating terms of like degree.

28. A polynomial in three variables can be written in one and only one way in the form $F_1(\sigma) + F_2(\sigma)x + F_3(\sigma)x^2 + F_4(\sigma)y + F_5(\sigma)xy + F_6(\sigma)x^2y$, where $F_i(\sigma)$ is a polynomial in $\sigma_1 = x + y + z, \sigma_2 = xy + yz + zx, \sigma_3 = xyz$. Write z^2 in this form. State and prove an analogous theorem for polynomials in n variables.

29. Let $F(X, r_1, r_2, \ldots, r_n)$ be a polynomial in $n + 1$ variables $X, r_1, r_2, \ldots, r_n$, say with integer coefficients. Show that if F has the property that substitution of r_i for X gives a polynomial that is identically 0 for $i = 1, 2, \ldots, n$ then $F = (X - r_1)(X - r_2) \cdots (X - r_n)Q$ for some polynomial Q in $X, r_1, r_2, \ldots, r_n$ with integer coefficients.

30. In connection with the proof of §11, show that an $n \times n$ system of linear equations $\sum a_{ij}x_j = y_i$ in which the a's, x's, and y's are polynomials in some fixed set of variables $r_1, r_2, \ldots, r_n$—or in fact where the a's, x's, and y's are elements of any commutative ring without zero divisors—determines the x's when the a's and y's are given unless $\det(a_{ij}) = 0$.

A Method for Solving the Cubic

§14 The solution of the quadratic equation is expressed very succinctly in the formula

$$x = \tfrac{1}{2}[(x + y) + (x - y)] = \tfrac{1}{2}[(x + y) + \sqrt{(x - y)^2}]. \tag{1}$$

Here x and y are supposed to be the roots of a quadratic equation $x^2 + bx + c = 0$, so that polynomials symmetric in x and y have known values expressible in terms of b and c; since $x + y$ and $(x - y)^2$ are symmetric polynomials, their values are known and can be substituted into (1) to give the value of x. The only hitch is that $\sqrt{(x - y)^2}$ has two values. If the other one, $y - x$, is used in (1), the result is the other root y instead of x. Thus, in either case, (1) gives a root of the equation and, since $x + y$ is known, the other root can then be found immediately. Since $x + y = -b$ and $(x - y)^2 = (x + y)^2 - 4xy = b^2 - 4c$, the solution in (1) is the familiar one $\tfrac{1}{2}(-b \pm \sqrt{b^2 - 4c})$.

A similar technique can be applied to the solution of the cubic. Let x, y, z be the solutions of a given cubic equation, so that symmetric polynomials in x, y, z have known values expressible in terms of the coefficients of the given cubic. In a solution of the cubic analogous to (1), it is natural to expect that cube roots will be involved rather than square roots. In the same way that a quantity u has two square roots $\pm\sqrt{u}$, it has three cube roots $\sqrt[3]{u}, \alpha(\sqrt[3]{u}), \alpha^2(\sqrt[3]{u})$, where $\sqrt[3]{u}$ is one of the cube roots (which is normally taken to be the real cube root if u is real) and α is a primitive cube root of 1, that is, $\alpha^3 = 1, \alpha \neq 1$. There are two possible values for α and they satisfy the quadratic equation $\alpha^2 + \alpha + 1 = (\alpha^3 - 1)/(\alpha - 1) = 0$, which, by the quadratic formula, gives $\alpha = (-1 \pm \sqrt{-3})/2$.

The analog of (1) for the solution of the cubic is

$$x = \tfrac{1}{3}[(x + y + z) + (x + \alpha y + \alpha^2 z) + (x + \alpha^2 y + \alpha z)]$$
$$= \tfrac{1}{3}[(x + y + z) + \sqrt[3]{(x + \alpha y + \alpha^2 z)^3} + \sqrt[3]{(x + \alpha^2 y + \alpha z)^3}]. \tag{2}$$

This formula requires that the quantities under the radicals, say $u = (x + \alpha y + \alpha^2 z)^3$ and $v = (x + \alpha^2 y + \alpha z)^3$, be evaluated. This is easily done by observing that uv and $u + v$ are symmetric in x, y, z and are therefore known. Clearly uv and $u + v$ are invariant under interchange of y and z because this operation merely interchanges u and v. They are invariant under the cyclic permutation $x \to z \to y \to x$ because this carries u to

$$(z + \alpha x + \alpha^2 y)^3 = [(\alpha)(\alpha^2 z + x + \alpha y)]^3 = u$$

and, in the same way, carries v to v. Since any permutation of x, y, z can be obtained as a composition of interchanges of y and z and cyclic permutations, it follows that uv and $u + v$ are symmetric polynomials in x, y, z and can be evaluated. (See Exercise 1.) Thus u and v can be evaluated using (1). Once this is done, these values can be used in (2) to find x. One has no way of telling which is u and which is v, but since u and v enter symmetrically in (2) this is no problem. What is a problem in using (2) is the choice of the cube roots of u and v. Since there are three choices for each, there are nine values of (2). Of these, two more,

$$y = \tfrac{1}{3}[(x + y + z) + \alpha^2(x + \alpha y + \alpha^2 z) + \alpha(x + \alpha^2 y + \alpha z)],$$

and

$$z = \tfrac{1}{3}[(x + y + z) + \alpha(x + \alpha y + \alpha^2 z) + \alpha^2(x + \alpha^2 y + \alpha z)],$$

are the remaining solutions of the cubic, while the remaining six are not solutions. One can thus use (2) to find nine values and then simply try all nine in the cubic and reject those which fail to satisfy it. (For a better method of selecting the three solutions from the nine values given by (2) see §16.)

(In this solution we have assumed, as did the mathematicians of the eighteenth century, that the roots x, y, z *exist* in some sense and that the problem is merely to give formulas for them. The problem of the existence of roots is discussed beginning in §49.)

Lagrange (Vandermonde) Resolvents

§15 The solution of the cubic equation in the preceding article was first given by the French mathematician Vandermonde in a presentation to the Paris Academy in 1770 [V1]. Vandermonde went beyond the cubic to the quartic and to certain higher degree equations, including the equation for 11th roots of unity (see §22). Unfortunately, although Vandermonde was still young at the time, and although this work demonstrated a great deal of insight into the theory of equations, this is as far as his work went. It was eclipsed a few months later by the publication of Lagrange's extensive *Réflexions sur la Résolution Algébrique des Equations* by the Prussian Academy while Vandermonde's work was still awaiting publication by the Paris Academy. (It waited until 1774.)

(For an interesting account of Vandermonde's career, see Lebesgue's essay [L2]. Although Vandermonde had insightful ideas in other areas of mathematics as well, he does not seem to have followed these up either, and he is remembered today only because the name "Vandermonde determinant" was given to a determinant which, ironically, does not occur in his work at all.*)

---

* Lebesgue conjectures that this name was concocted by someone who misread a passage in Vandermonde, mistaking superscripts for exponents. However, the formula $a^2b + b^2c + c^2a - a^2c - b^2a - c^2b = (a - b)(a - c)(b - c)$ does occur early in Vandermonde's main treatise. If the left side is written as a 3×3 determinant

$$\begin{vmatrix} a^2 & b^2 & c^2 \\ a & b & c \\ 1 & 1 & 1 \end{vmatrix},$$

it is a "Vandermonde determinant" and the generalization to four or more variables is easy to derive.

Unlike Vandermonde, who was French but did not have a French name, Lagrange had a French name but was not French. He was Italian (he was born with the name Lagrangia and his native city was Turin) and at the time of the publication of his *Réflexions* he was a member of Frederick the Great's Academy in Berlin. When he left Berlin in 1787 he went to Paris, where he spent the rest of his life and where, of course, he was a leading member of the scientific community; this has tended to reinforce the impression that he was French. He was certainly the greatest mathematician of the generation between Euler and Gauss, and, indeed, has a secure place among the great mathematicians of all time.

§16   As was mentioned in §2, Lagrange's *Réflexions* was a very long work (in his *Oeuvres* it occupies 220 pages) that examines from all angles the solution of equations of degree 2, 3, and 4, and seeks to deal with the algebraic solution of equations of higher degree. His approach to the solution of the cubic is essentially Vandermonde's (though Lagrange was not aware of Vandermonde's work at the time) but he describes it in a somewhat different way. For him, as for Vandermonde, the basic idea is to consider the quantity $x + \alpha y + \alpha^2 z$ where x, y, z are the solutions of the cubic in question and where α is a cube root of unity ($\alpha \neq 1$). Let t denote this quantity. Lagrange observes that t has six values, depending on the order in which the roots x, y, z are taken. These six values are the solutions of a 6th degree equation

$$f(X) = (X - t_1)(X - t_2)(X - t_3)(X - t_4)(X - t_5)(X - t_6) = 0,$$

whose coefficients, being symmetric in the six values of t, are symmetric in x, y, z and are therefore known quantities expressible in terms of the coefficients of the given cubic. Lagrange calls this the *resolvent* equation. Although it is of higher degree than the original, it is *solvable* because it is in fact a *quadratic equation in* X^3 and can be solved by solving a quadratic and then taking a cube root.

The fact that the resolvent equation $f(X) = 0$ is quadratic in X^3 is easily seen by observing that the values of t can be ordered so that $t_2 = z + \alpha x + \alpha^2 y = \alpha t_1$, $t_3 = \alpha^2 t_1$, $t_4 = x + \alpha z + \alpha^2 y$, $t_5 = \alpha t_4$, $t_6 = \alpha^2 t_4$, which gives $(X - t_1)(X - t_2)(X - t_3) = (X - t_1)(X - \alpha t_1)(X - \alpha^2 t_1) = X^3 - t_1^3$ and $f(X) = (X^3 - t_1^3)(X^3 - t_4^3) = X^6 - (t_1^3 + t_4^3)X^3 + t_1^3 t_4^3$. In the notation above, $u = t_1^3$, $v = t_4^3$, and the coefficients of $f(X)$ are precisely the quantities $u + v$ and uv whose expression in terms of the coefficients of the original cubic gave the solution above. Once the six values of t are obtained from the solution of the resolvent equation, the solutions of the cubic are given by $x = \frac{1}{3}[(x + y + z) + t_1 + t_4]$, $y = \frac{1}{3}[(x + y + z) + \alpha^2 t_1 + \alpha t_4]$, $z = \frac{1}{3}[(x + y + z) + \alpha t_1 + \alpha^2 t_4]$ and the only problem is to identify t_1 and t_4 among the six solutions of the resolvent equation.

Unlike Vandermonde, Lagrange deals with the problem of determining which t's to use in the formulas for x, y, z. Let t be any solution of the resolvent equation. Then the roots x, y, z can be reordered, if necessary, so that $t = x + \alpha y + \alpha^2 z$. Lagrange observes that $(x + \alpha y + \alpha^2 z)(x + \alpha^2 y + \alpha z)$ *is symmetric in* x, y, z *and is therefore a known quantity, say* w. (See Exercise 3.) Once this observation has been made, the solution is simple. The three roots of the cubic are $\frac{1}{3}[(x + y + z) + t + (w/t)]$, $\frac{1}{3}[(x + y + z) + \alpha t + (w/\alpha t)]$, and $\frac{1}{3}[(x + y + z) + \alpha^2 t + (w/\alpha^2 t)]$, where t is any one of the six solutions of the resolvent equation.

§17 The natural way to generalize Lagrange's method to the solution of quartic equations is to set $t = x + iy - z - ir$ $(= x + iy + i^2 z + i^3 r)$ where $i = \sqrt{-1}$ is a primitive* 4th root of unity and x, y, z, r are the roots of the quartic in question. Then t has $4! = 24$ different values according to the order of x, y, z, r, and they satisfy a resolvent equation $f(X) = 0$ of degree 24. Can the resolvent equation be solved and, if so, does its solution lead to a solution of the original equation? The answer to these two questions, as both Vandermonde and Lagrange showed, is yes (see Exercise 4), but, as both of them also noted, there is an easier way to derive a solution of the quartic from considerations of this type.

The analog of formula (2) of §14 would be

$$x = \tfrac{1}{4}[(x + y + z + r) + \sqrt[4]{(x + iy - z - ir)^4} + \sqrt[4]{(x - y + z - r)^4}$$
$$+ \sqrt[4]{(x - iy - z + ir)^4}].$$

To use this it would be necessary to evaluate each of the quantities under the radical signs. In particular, it would be necessary to evaluate the middle one $(x - y + z - r)^4$. But, in fact, the evaluation of this one quantity, which is relatively easy using ideas that have already been introduced, is enough to make possible the solution of the quartic equation as follows. Let $t = x - y + z - r$. Then the twenty-four permutations of x, y, z, r produce only six different values of t, namely, $\pm(x - y + z - r)$, $\pm(x + y - z - r)$, $\pm(x - y - z + r)$, and each occurs four times. Let them be denoted by $\pm t_1$, $\pm t_2$, $\pm t_3$. Then the resolvent equation associated with t is of the form

$$f(X) = (X - t_1)^4(X + t_1)^4(X - t_2)^4(X + t_2)^4(X - t_3)^4(X + t_3)^4$$

$$= g(X)^4 = 0$$

where $g(X) = (X^2 - t_1^2)(X^2 - t_2^2)(X^2 - t_3^2)$. Thus t_1^2, t_2^2, t_3^2 are the roots of a *known* cubic equation, because the coefficients of $g(X)$ are symmetric in t_1^2, t_2^2, t_3^2 and therefore symmetric in x, y, z, r. Since a cubic equation can be solved, this means that t_1^2, t_2^2, t_3^2 can be found and therefore, by taking square

---

* An nth root of unity α is said to be primitive if n is the *least* positive integer such that $\alpha^n = 1$. Of the four 4th roots of unity (± 1, $\pm i$) two are primitive ($\pm i$).

roots, so can $\pm t_1$, $\pm t_2$, $\pm t_3$. Since

$$x = \tfrac{1}{4}[(x + y + z + r) + t_1 + t_2 + t_3],$$

$$y = \tfrac{1}{4}[(x + y + z + r) - t_1 + t_2 - t_3],$$

$$z = \tfrac{1}{4}[(x + y + z + r) + t_1 - t_2 - t_3],$$

$$r = \tfrac{1}{4}[(x + y + z + r) - t_1 - t_2 + t_3],$$

the solution of the quartic is reduced to the problem of assigning signs* to t_1, t_2, t_3. Let s_1 be any one of the six roots of $g(X) = 0$, and let s_2 be any other root with† $s_2 \neq -s_1$. Then a check of the possibilities shows that just one root of the quartic occurs with the sign $+$ in both s_1 and s_2, and just one with the sign $-$ in both. Rename the roots so that x occurs with the sign $+$ in both s_1 and s_2 and r with the sign $-$ in both. Of the remaining two roots, let y be the one which occurs with the sign $-$ in s_1 and let z be the last root. With the roots so named, $s_1 = t_1$ and $s_2 = t_2$. In short, any two distinct solutions of $g(X) = 0$ can be taken to be t_1 and t_2, provided only that one is not the negative of the other. The above formulas can then be used to find x, y, z, and r once t_3 is determined. This can be done simply by observing that $t_1 t_2 t_3$ is symmetric in x, y, z, r and is therefore known, say $t_1 t_2 t_3 = w$. (See Exercise 5.) Then $t_3 = w/t_1 t_2$ is determined by t_1 and t_2, which completes the solution of the quartic.

§18   Consider now the extension of these ideas to the case of a quintic equation. Let x_1, x_2, x_3, x_4, x_5 be the roots of the equation and let $t = x_1 + \alpha x_2 + \alpha^2 x_3 + \alpha^3 x_4 + \alpha^4 x_5$ where α is a primitive 5th root of unity, $\alpha^5 = 1, \alpha \neq 1$. Then t has $5! = 120$ different values and it satisfies an equation $f(X) = 0$ of degree 120 whose coefficients are symmetric polynomials in the x's and are therefore known. If t_1 is a value of t then so are $\alpha t_1 = x_5 + \alpha x_1 + \alpha^2 x_2 + \dots$, and $\alpha^2 t_1$, $\alpha^3 t_1$, $\alpha^4 t_1$. Thus $f(X)$ contains the factor‡ $(X - t_1)(X - \alpha t_1)(X - \alpha^2 t_1)(X - \alpha^3 t_1)(X - \alpha^4 t_1) = X^5 - t_1^5$ and   is, actually, the product of twenty-four such factors. Thus, although $f(X)$ has degree 120, it is in fact a polynomial of degree 24 in X^5. Is it possible to solve $f(X) = 0$ and, if so, can one use the solution to find x_1, x_2, x_3, x_4, x_5?

Since it is easy (see Exercise 7) to find values t_1, t_2, t_3, t_4 of t such that $x_1 = \tfrac{1}{5}[(x_1 + x_2 + x_3 + x_4 + x_5) + t_1 + t_2 + t_3 + t_4]$, the answer to the second question is "yes" if one is prepared to resort to a good deal of trial

---

* As with the cubic, Vandemonde did not address this problem and contented himself with the formula $\tfrac{1}{4}[(x + y + z + r) \pm t_1 \pm t_2 \pm t_3]$ which gives eight values, four of which are roots of the equation.

† This is possible unless two of the roots of the original quartic are equal (see Exercise 8), a case which can be ignored for the reason explained in §31.

‡ This formula makes use of the equation $(x - 1)(x - \alpha)(x - \alpha^2)(x - \alpha^3)(x - \alpha^4) = x^5 - 1$, which follows from the fact that $1, \alpha, \alpha^2, \alpha^3, \alpha^4$ are the roots of $x^5 - 1$.

and error. Perhaps it would be possible to reduce the trial and error or even to eliminate it altogether, as was possible in the cases of the cubic and the quartic, but there is little point in pursuing this question unless there is some hope that the answer to the first question is "yes." Neither Lagrange nor Vandermonde thought this was very likely.

What is happening, clearly, is that the degree of the resolvent equation $f(X) = 0$ is growing too rapidly. When $n = 3$ the resolvent equation has degree $3! = 6$ but actually has degree $2! = 2$ in X^3 and is therefore solvable. When $n = 4$ it has degree $4! = 24$ but actually has degree $3! = 6$ in X^4. With ingenuity this particular equation of degree 6 can be solved. Alternatively, if, in the case $n = 4$, the resolvent $t = x_1 + \alpha x_2 + \alpha^2 x_3 + \alpha^3 x_4$ with $\alpha = -1$ instead of the *primitive* 4th root $\alpha = i$ is used, then the resolvent equation $f(X) = g(X)^4$ is the 4th power of a 6th degree equation $g(X)$ which is easily solvable because it is actually a cubic in X^2. This trick cannot be used in the case $n = 5$ because all 5th roots of unity other than 1 are primitive. Thus, in the case $n = 5$ one is faced with the solution of an equation of degree 24 and, although it is a special equation, this seems an extremely difficult and possibly hopeless task. At least Lagrange thought so.

§19   This is the essence of Lagrange's *Réflexions* as far as the solution of equations of low degree by radicals is concerned. He examines in detail all the approaches that have been taken to the solution of cubic and quartic equations and shows how they can all be interpreted as applications of one method. A quantity t, called the *resolvent*, is obtained as the solution of an auxiliary equation called the *resolvent equation*, and the roots of the original equation are expressed in terms of t. Moreover, he shows that in the successful cases the resolvent has the form $x_1 + \alpha x_2 + \cdots + \alpha^{n-1} x_n$ where n is the degree of the equation, where the x_i are the roots of the equation, and where α is an nth root of unity (not necessarily primitive). Such a quantity is now known as a *Lagrange resolvent*. After remarking on the apparent failure of the method for equations of degree 5 or greater, he proceeds to a very general investigation of possible resolvents t in terms of which one might hope to express the roots of the equation. This investigation leads to no particular conclusion—although it includes a general theorem of considerable importance (see §29)—and leaves the subject in an unfinished condition well designed to invite a young Galois to attempt to carry it further.

Second Exercise Set

1. Show that the six permutations of x, y, z can be obtained as compositions of the cyclic permutation $x \to y \to z \to x$ and the interchange $y \leftrightarrow z$.

2. Let u and v be as in §14 and find uv and $u + v$ in terms of $\sigma_1 = x + y + z$, $\sigma_2 = xy + yz + zx$, and $\sigma_3 = xyz$.

3. Find $(x + \alpha y + \alpha^2 z)(x + \alpha^2 y + \alpha z)$ in terms of $\sigma_1, \sigma_2, \sigma_3$.

4. Fill out the following sketch of a solution of the quartic equation based on the resolvent $t = x + iy - z - ir$. (See Lagrange [L1], Section 48. Lagrange ascribes the method to Bezout.) This quantity satisfies an equation of degree 24 which is an equation of degree 6 in t^4. Let $\pi = t\bar{t}$ where $\bar{t}$ is the complex conjugate of t. Then π is the root of a known cubic. Then $(i/2)(t + \bar{t})(t - \bar{t})$ is the difference of the other two roots of this cubic, and its square $-(t^4 - 2t^2\bar{t}^2 + \bar{t}^4)/4$ can be expressed in terms of π. Thus t satisfies a quadratic equation in t^4 whose coefficients are expressible in terms of π. A solution x of the original equation can be expressed in terms of t, $\bar{t}$, a known quantity, and $x - y + z - r$. The latter can be found either as in §17 or in terms of other values of t.

5. Find the expression of $t_1 t_2 t_3 = w$ in terms of $\sigma_1, \sigma_2, \sigma_3, \sigma_4$ where, as in §17, $t_1 = x - y + z - r, t_2 = x + y - z - r, t_3 = x - y - z + r$.

6. In the notation of Exercise 5, show that if $t_1 = \pm t_2$ or $\pm t_3$ then two roots of the original equation are equal.

7. Find four values t_1, t_2, t_3, t_4 of $t = x_1 + \alpha x_2 + \alpha^2 x_3 + \alpha^3 x_4 + \alpha^4 x_5$, as in §18, such that $x_1 = \frac{1}{5}[\sigma_1 + t_1 + t_2 + t_3 + t_4]$.

8. Show that the six quantities

$$\pm(x - y + z - r), \quad \pm(x + y - z - r), \quad \pm(x - y - z + r)$$

contain two, say s and t, such that $s \neq \pm t$, unless at least three of x, y, z, r are equal.

Cyclotomic Equations

§20 Roots of unity are of great importance in algebra both because they occur in connection with taking any root (if $\sqrt[n]{c}$ is any nth root of c then the most general nth root of c is $\alpha\sqrt[n]{c}$ where α is an nth root of unity) and because they occur in the Lagrange resolvent.

Roots of unity can be described *transcendentally* by the formula $\alpha = \cos(2\pi k/n) + i\sin(2\pi k/n)$ (where $k = 1, 2, \ldots, n$) because de Moivre's formula $(\cos\theta + i\sin\theta)^n = \cos n\theta + i\sin n\theta$ then gives $\alpha^n = \cos(2\pi k) + i\sin(2\pi k) = 1$. In view of the geometric meaning of the circular functions $\sin x$, $\cos x$, this means that the roots of unity are the vertices of a regular n-gon inscribed in the unit circle $\{|z| = 1\}$ of the complex z-plane with one vertex of the n-gon at $z = 1$. For this reason, the algebraic equation $x^n = 1$ satisfied by the roots of unity is called the *cyclotomic equation*—its solution effects the division of a circle into n equal parts (cycle = circle, tom = division or part).

One of the basic problems in the theory of the algebraic solution of equations is to give *algebraic* solutions of the cyclotomic equations $x^n = 1$.

§21 In the case $n = 3$, the algebraic solution

$$\alpha^2 + \alpha + 1 = (\alpha^3 - 1)/(\alpha - 1) = 0,$$

$\alpha = \frac{1}{2}(-1 \pm \sqrt{-3})$ was given above in §14. (The same solution can easily be found geometrically as well. See Exercise 1.) In the case $n = 4$, the algebraic solution $\alpha^2 + 1 = (\alpha^4 - 1)/(\alpha^2 - 1) = 0$, $\alpha = \sqrt{-1}$ is even easier. In the case $n = 5$, a primitive root α satisfies $\alpha^4 + \alpha^3 + \alpha^2 + \alpha + 1 = (\alpha^5 - 1)/(\alpha - 1) = 0$. Thus it satisfies a quartic equation and an algebraic solution is possible. However, instead of applying the general method of solving quartic equations, it is easier to exploit the symmetry of the equation $\alpha^4 + \alpha^3 + \alpha^2 + \alpha + 1 = 0$ to reduce it to a quadratic. This can be done by dividing by α^2 and noting that $(\alpha + \alpha^{-1})^2 = \alpha^2 + 2 + \alpha^{-2}$ to write the equation in the form $(\alpha + \alpha^{-1})^2 - 2 + (\alpha + \alpha^{-1}) + 1 = 0$, $\gamma^2 + \gamma - 1 = 0$, where $\gamma = \alpha + \alpha^{-1}$. Then $\gamma = \frac{1}{2}(-1 \pm \sqrt{5})$ and $\alpha^2 - \alpha\gamma + 1 = 0$, $\alpha = \frac{1}{2}(\gamma \pm \sqrt{\gamma^2 - 4})$, from which it follows (see Exercise 3) that

$$\alpha = \frac{\sqrt{5} - 1 \pm \sqrt{-2\sqrt{5} - 10}}{4}, \qquad \frac{-\sqrt{5} - 1 \pm \sqrt{2\sqrt{5} - 10}}{4}$$

are the four primitive 5th roots of unity. (The algebraic fact that the 5th roots of unity can be expressed in terms of square roots corresponds to the geometrical fact that a regular pentagon can be constructed by ruler and compass, a fact which is a major feature of Euclid's *Elements*. See Exercise 2.)

If $n = jk$ where j and k are relatively prime, then it is easy to see (Exercise 4) that the product of a primitive jth root and a primitive kth root of unity is a primitive nth root of unity. Thus, $(-1)\alpha$ is a primitive 6th root of unity, where α is a primitive cube root of unity. More generally, since any n can be written in the form $n = j \cdot k \cdot \cdots \cdot m$ where the factors $j, k, \ldots, m$ are *prime powers* and are relatively prime, this fact shows that in order to be able to find primitive nth roots of unity for an arbitrary positive integer n, it suffices to be able to do this in the case where n is a *prime power*. The principal case is the case where n is *prime*, and this is the case that will be treated here. It is then simple to show (see Exercise 15) that the p^{n-1}st root of a primitive pth root of unity is a primitive p^nth root of unity.

Eleventh Roots of Unity

§22 For the next prime, $n = 7$, the trick used in the case $n = 5$ can be imitated to give $\alpha^6 + \alpha^5 + \alpha^4 + \alpha^3 + \alpha^2 + \alpha + 1 = (\alpha^7 - 1)/(\alpha - 1) = 0$, $\alpha^3 + \alpha^{-3} + \alpha^2 + \alpha^{-2} + \alpha + \alpha^{-1} + 1 = 0$, $\gamma^3 - 3\gamma + \gamma^2 - 2 + \gamma + 1 = 0$, $\gamma^3 + \gamma^2 - 2\gamma - 1 = 0$, where $\gamma = \alpha + \alpha^{-1}$. The cubic equation for γ can be solved (see Exercise 11) and the result used in $\alpha^2 - \alpha\gamma + 1 = 0$, $\alpha = \frac{1}{2}(\gamma \pm \sqrt{\gamma^2 - 4})$ to give α.

When this reduction technique is applied in the case $n = 11$, it reduces $(\alpha^{11} - 1)/(\alpha - 1) = \alpha^{10} + \alpha^9 + \cdots + \alpha + 1 = 0$ to a *quintic* equation and therefore does *not* in itself lead to an algebraic solution of the equation.

In his *Réflexions*, Lagrange observes that the solution of $x^{11} = 1$ leads to a quintic equation, and leaves it at that. Vandermonde, on the other hand, gave an algebraic solution of this problem, and, in this, far excelled Lagrange. His method can be briefly described as an application of the Lagrange (Vandermonde) resolvent to the quintic that comes from the above reduction. The essence of his idea can be seen more clearly, however, if one applies it not to the quintic equation but to the original 10th degree equation

$$\alpha^{10} + \alpha^9 + \cdots + \alpha + 1 = (\alpha^{11} - 1)/(\alpha - 1) = 0.$$

In order to apply the method of the Lagrange resolvent to a 10th degree equation, one needs a 10th root of unity. Let β be a primitive 10th root of unity (that is, β is the negative of one of the primitive 5th roots found above) and let α be the desired 11th root of unity. Then the roots of the equation $x^{10} + x^9 + \cdots + x + 1 = 0$ are $\alpha, \alpha^2, \alpha^3, \ldots, \alpha^{10}$, so the Lagrange resolvent is $\alpha^j + \beta\alpha^k + \cdots + \beta^9 \alpha^m$, where $(j, k, \ldots, m)$ is some permutation of $(1, 2, \ldots, 10)$. The basic idea* behind the solution that follows is to put the roots of $(x^{11} - 1)/(x - 1)$ in the order $\alpha, \alpha^2, \alpha^4, \alpha^8, \alpha^5, \alpha^{10}, \alpha^9, \alpha^7, \alpha^3, \alpha^6$ in which each root is the *square* of its predecessor. The associated Lagrange resolvent is then

$$t = \alpha + \beta\alpha^2 + \beta^2\alpha^4 + \beta^3\alpha^8 + \beta^4\alpha^5 + \beta^5\alpha^{10} + \beta^6\alpha^9 \\ + \beta^7\alpha^7 + \beta^8\alpha^3 + \beta^9\alpha^6. \tag{1}$$

The advantage of putting the roots in this order is that t^{10} is then a known quantity, as will be shown below. More generally, let t_i, for $i = 1, 2, \ldots, 10$, be t with β changed to β^i. (In particular, $t_{10} = \alpha + \alpha^2 + \alpha^4 + \alpha^8 + \cdots + \alpha^6 = -1$.) Then t_i^{10} is known for all i and $\alpha = \frac{1}{10}[t_1 + t_2 + \cdots + t_{10}] = \frac{1}{10}[\sqrt[10]{t_1^{10}} + \sqrt[10]{t_2^{10}} + \cdots + \sqrt[10]{t_{10}^{10}}]$, so that, except for the problem of choosing the 10th roots, a primitive 11th root α has been found. Actually there is no problem at all in choosing the 10th roots because the same argument that shows that t_i^{10} is a known quantity also shows that $t_i t_1^{10-i}$ is a known quantity so that once t_1 has been chosen (and there are just ten possible choices, because t_1^{10} is known) the values of $t_2, t_3, \ldots, t_9$ are known and therefore so is $\alpha = \frac{1}{10}(t_1 + t_2 + \cdots + t_{10})$. This procedure is easily seen to give an 11th root of unity α no matter which of the 10th roots of t_1^{10} is used (see Exercise 5).

§23 The main step in this solution of $x^{11} = 1$ is to show that if t is defined by (1) of §22 then t^{10} is a known quantity. What this means, in essence, is that t^{10} can be expressed in terms of the known quantity β alone.

---

* In Lebesgue's interpretation, Vandermonde was aware of the connection between the order he chose for the roots of the quintic and the fact that 2 is a primitive root mod 11. This is probable but by no means certain. See Lebesgue [L2].

Now t^{10} is, in the first instance, a polynomial in α and β. Thus, a typical term of t^{10} is of the form $A\beta^j\alpha^k$ where A, j, and k are nonnegative integers. Since $\beta^{10} = 1$ and $\alpha^{11} = 1$, one can assume $0 \le j \le 9$ and $0 \le k \le 10$. Now let the terms with like powers of α be gathered to give an expression of the form

$$
\begin{aligned}
t^{10} = {} & p_0(\beta) + p_1(\beta)\alpha + p_2(\beta)\alpha^2 + p_3(\beta)\alpha^4 \\
& + p_4(\beta)\alpha^8 + p_5(\beta)\alpha^5 + p_6(\beta)\alpha^{10} \\
& + \cdots + p_{10}(\beta)\alpha^6,
\end{aligned}
\tag{2}
$$

where $p_i(\beta)$ is a polynomial of degree < 10 in β with integer coefficients. What is to be shown is that t^{10} does not in fact involve α.

The key fact is that *changing α to α^2 changes t to $\beta^{-1}t$ and therefore leaves t^{10} unchanged*. These two statements are immediate consequences of the definitions of t and β. Therefore

$$
\begin{aligned}
& p_0(\beta) + p_1(\beta)\alpha + p_2(\beta)\alpha^2 + p_3(\beta)\alpha^4 + \cdots + p_{10}(\beta)\alpha^6 \\
& = p_0(\beta) + p_1(\beta)\alpha^2 + p_2(\beta)\alpha^4 + p_3(\beta)\alpha^8 + \cdots + p_{10}(\beta)\alpha,
\end{aligned}
$$

from which

$$
\begin{aligned}
0 = {} & [p_1(\beta) - p_{10}(\beta)]\alpha + [p_2(\beta) - p_1(\beta)]\alpha^2 \\
& + [p_3(\beta) - p_2(\beta)]\alpha^4 + \cdots + [p_{10}(\beta) - p_9(\beta)]\alpha^6.
\end{aligned}
$$

An equation of the form $0 = q_1(\beta)\alpha + q_2(\beta)\alpha^2 + q_3(\beta)\alpha^4 + \cdots + q_{10}(\beta)\alpha^6$ implies that $q_1(\beta) = 0$, $q_2(\beta) = 0, \ldots, q_{10}(\beta) = 0$. (See Lemma 2 below.) Therefore in (2) we have $p_1(\beta) = p_2(\beta) = p_3(\beta) = \cdots = p_{10}(\beta)$. Let $p(\beta)$ denote their common value. Then $t^{10} = p_0(\beta) + p(\beta)(\alpha + \alpha^2 + \alpha^4 + \cdots + \alpha^6) = p_0(\beta) + p(\beta)(-1) = p_0(\beta) - p(\beta)$ is independent of α, as required. The proof that $t_i t_1^{10-i}$ is independent of α follows in exactly the same way from the fact that changing α to α^2 multiplies t_i by β^{-i} so that $t_i t_1^{10-i}$ is multiplied by $\beta^{-i}(\beta^{-1})^{10-i} = \beta^{-10} = 1$. Finally, the same Lemma 2 shows (because $t = 0$ would imply $\beta = 0$, $\beta^2 = 0$, $\beta^3 = 0, \cdots$) that $t \ne 0$, which is needed in the determination of t_i once $t_i t_1^{10-i}$ is known. This technique reduces the construction not only of an 11th root of unity but also of a pth root of unity for any prime $p > 2$ to the two lemmas of the following article.

The Cases $p > 11$

§24   **Lemma 1.** *For every prime p there is an integer g with the property that every integer not congruent to 0 mod p is congruent to a power of g mod p. Such an integer g is called a* primitive root mod p.

If α is a primitive pth root of unity and if g is a primitive root mod p then $\alpha, \alpha^g, \alpha^{g^2}, \alpha^{g^3}, \ldots, \alpha^{g^{p-2}}$ is a list of all the primitive pth roots of unity and, in analogy with (1) of §22, one can set $t = \alpha + \beta\alpha^g + \beta^2\alpha^{gg} + \cdots$ where β is a primitive $(p - 1)$st root of unity. Then, exactly as in the case $p = 11$, t

changes to $\beta^{-1}t$ when α is changed to α^g; thus t^{p-1} and $t_i t^{p-1-i}$ are unchanged by $\alpha \mapsto \alpha^g$ when t_i is t with β^i in place of β. Thus

$$\alpha = (p-1)^{-1}(t_1 + t_2 + \cdots + t_{p-1})$$

is a description of α in terms of known quantities and the $(p-1)$st root of a known quantity once Lemma 2 is proved.

Lemma 2. *Let p be a prime, let α be a primitive pth root of unity, and let β be a primitive $(p-1)$st root of unity. If $P_1(\beta), P_2(\beta), \ldots, P_{p-1}(\beta)$ are polynomials in β with rational coefficients and if $P_1(\beta)\alpha + P_2(\beta)\alpha^2 + \cdots + P_{p-1}(\beta)\alpha^{p-1} = 0$ then $P_1(\beta) = 0, P_2(\beta) = 0, \ldots, P_{p-1}(\beta) = 0$.*

When t^{p-1} is written in the form $p_0(\beta) + p_1(\beta)\alpha + p_2(\beta)\alpha^g + p_3(\beta)\alpha^{gg} + \cdots$, the fact that it is invariant under $\alpha \mapsto \alpha^g$ shows that it is equal to $p_0(\beta) + p_1(\beta)\alpha^g + p_2(\beta)\alpha^{gg} + \cdots$, and subtraction of these equal quantities gives $0 = [p_1(\beta) - p_{p-1}(\beta)]\alpha + [p_2(\beta) - p_1(\beta)]\alpha^g + \cdots$. Then, by Lemma 2,

$$p_1(\beta) = p_2(\beta) = \cdots = p_{p-1}(\beta)$$

and

$$t^{p-1} = p_0(\beta) + p_1(\beta)(\alpha + \alpha^2 + \alpha^3 + \cdots + \alpha^{p-1}) = p_0(\beta) - p_1(\beta)$$

is independent of α and can therefore be regarded as known. The $(p-1)$st root of this known quantity can then be set equal to t. Since the same argument shows that $t_i t^{p-1-i}$ is known, and that $t \neq 0$, it follows that t_i can be expressed in terms of t and known quantities. Therefore

$$\alpha = (p-1)^{-1}(t_1 + t_2 + t_3 + \cdots + t_{p-1})$$

has been found algebraically.

These two lemmas are of very different orders of difficulty. The first is not hard. (See Exercise 7.) It was evidently well known to be true in the eighteenth century, but, according to Gauss, his proof of it in Article 55 of the *Disquisitiones Arithmeticae* was the first rigorous proof of it to be published.* The second, on the other hand, is rather hard. Gauss strongly implied in the *Disquisitiones* that he had a proof, but he did not give one.† For a proof, see §71.

---

* See, however, p. XXIX of volume 3 of Euler's *Opera*.

† Gauss's statements about Lemma 2 are rather mystifying. At one point he seems to use a generalization of Lemma 2 without proof (Art. 360, I, when he equates corresponding coefficients in U and U'). At another point, he omits the proof that $t \neq 0$, saying that it would be too long (Art. 360, III). However, as was noted above, $t \neq 0$ is an immediate consequence of Lemma 2. Because of this gap in Gauss's proof, it appears to me correct to say that the equation $(x^p - 1)/(x - 1) = 0$ for pth roots of unity had not been shown to be solvable by radicals prior to Galois' work.

§25   The above method of constructing a pth root of unity α (when p is prime) was given by Gauss in the final section of the *Disquisitiones Arithmeticae*. This construction was not Gauss's main objective, however. Rather, he was interested in reducing the solution of the equation

$$x^{p-1} + x^{p-2} + \cdots + x + 1 = 0$$

satisfied by α to the solution of a *succession* of equations of lower degree. It will not be necessary to study his results in detail (see Exercises 8–13) but the basic idea is simple enough to understand and is illuminating.

Let p be a given prime ($p > 2$) and let α be a primitive pth root of unity. Let $\mathbb{Q}(\alpha)$ denote the field* of numbers of the form

$$a_0 + a_1\alpha + a_2\alpha^2 + \cdots + a_k\alpha^k$$

where $a_0, a_1, \ldots, a_k$ are rational numbers. (Here k is an arbitrarily large positive integer, but, since $\alpha^p = 1$, there is no point in taking k larger than $p - 1$.) If α can be expressed algebraically in terms of known quantities—that is, expressed by an algebraic formula involving addition, subtraction, multiplication, division, and the extraction of roots of known quantities†—then so can all elements of $\mathbb{Q}(\alpha)$. Conversely, since α is an element of $\mathbb{Q}(\alpha)$, if all elements of $\mathbb{Q}(\alpha)$ can be expressed algebraically then of course α can. In short, α can be expressed algebraically if and only if all elements of $\mathbb{Q}(\alpha)$ can. Now of course the elements of the field of rational numbers $\mathbb{Q}$ can be expressed algebraically in terms of the basic operations and the number 1. (In fact, elements of $\mathbb{Q}$ can be expressed *rationally* in terms of 1, that is, without the extraction of roots.) Gauss's technique for the solution of the problem was to construct a sequence of *intermediate* fields between the known field $\mathbb{Q}$ and the unknown field $\mathbb{Q}(\alpha)$ and to regard the whole extension from $\mathbb{Q}$ to $\mathbb{Q}(\alpha)$ as being made up of a sequence of extensions in which one ascends the ladder of intermediate extensions.

Specifically, let g be a primitive root mod p and let $S: \mathbb{Q}(\alpha) \rightarrow \mathbb{Q}(\alpha)$ be the operation of replacing α by α^g. (Since α^g, like α, is a primitive pth root of unity, S is in fact a well-defined automorphism of the field $\mathbb{Q}(\alpha)$.) By Lemma 2 of §24, the only elements of $\mathbb{Q}(\alpha)$ that are left fixed by S are the rational numbers. ‡ Let d be any divisor of $p - 1$ and let K_d be the set of all elements of $\mathbb{Q}(\alpha)$ that are invariant under S^d. As was just observed, $K_1 = \mathbb{Q}$. At the other extreme, since $g^{p-1} \equiv 1 \bmod p$, S^{p-1} is the identity operation and $K_{p-1} = \mathbb{Q}(\alpha)$. The intermediate fields mentioned above are the fields K_d (d a divisor of $p - 1$).**

---

* This is a technical term that will probably be familiar to most readers. It will be explained, for the benefit of readers to whom it is new, in §33.

† Since $\sqrt[p]{1} = 1$ is not a *primitive* pth root of unity, extraction of a pth root of the known quantity 1 does not solve the problem.

‡ This is in fact a special case of Lemma 2 which is much easier to prove than the full lemma. See Exercise 8.

** The elements $\alpha^i + S^d\alpha^i + S^{2d}\alpha^i + \cdots + S^{p-1-d}\alpha^i = \alpha^i + \alpha^{ig^d} + \alpha^{ig^{2d}} + \cdots + \alpha^{ig^{p-1-d}}$ of $\mathbb{Q}(\alpha)$ are what Gauss called "periods." Every element of K_d can be written as a linear combination of these periods with rational coefficients.

It is not hard to show that if d and D are divisors of $p - 1$ and if d divides D—say $D = qd$—then K_D is a simple algebraic extension of K_d of degree q. Roughly speaking, this means that the expression of elements of K_D in terms of elements of K_d involves the solution of an equation of degree q. (See §§34–36 for the definition of simple algebraic extensions.) The degree q of this equation is less than the degree $p - 1$ of the original equation for α (unless $d = 1$ and $D = p - 1$), and by interposing as many intermediate fields as possible Gauss was able to reduce the solution of the equation for α to the solution of a succession of equations of lower degree. If these equations of lower degree all have degrees less than 5 then they can be solved using the solution of equations of degree 2, 3, and 4 (Lagrange resolvents).

Gauss claimed that the particular equations that occur in the solution of the equation for α can always be solved by Lagrange resolvents (although he did not call them Lagrange resolvents) even when their degree is greater than 4. Specifically, he observed that the techniques applied above to the extension from $\mathbb{Q}$ to $\mathbb{Q}(\alpha)$ can also be applied (granted the appropriate analogue of Lemma 2) to the extension from K_d to K_D to prove:

Let d and D be divisors of $p - 1$ and let d divide D, say $D = qd$. Then any given element of K_D can be expressed rationally in terms of elements of K_d, the qth root of a particular element of K_d, and a primitive qth root of unity.

What is proved above is the case $d = 1, D = p - 1$, where the element to be expressed is α. In general, if γ is the given element of K_D, the key to the solution is the "Lagrange resolvent"

$$t = \gamma + \beta \cdot S^d \gamma + \beta^2 \cdot S^{2d} \gamma + \cdots + \beta^{q-1} \cdot S^{D-d} \gamma,$$

where β is a primitive qth root of unity. Then t^q is an element of K_d, so that t is the qth root of a known element. The expression of γ in terms of t and known elements is then easy to accomplish using adaptations of the techniques used in the previous case. (If γ itself is in K_d then $t = \gamma(1 + \beta + \cdots + \beta^{q-1}) = 0$, and the procedure does not work, but in this case γ is already known, by assumption. A crucial step of the proof, which Gauss omits, though he calls attention to it, is to show that $t = 0$ only when γ is contained in a proper subfield of K_D.)

This theorem, unlike the special case $d = 1, D = p - 1$, provides a construction of a primitive pth root of unity because it depends only on the knowledge of a primitive qth root of unity, and because one can *arrange the construction so that q is always a prime* (necessarily a divisor of $p - 1$). In fact, prime values of q arise when one interposes, as it is natural to do, the maximum number of fields K_d between $K_{p-1} = \mathbb{Q}(\alpha)$ and $K_1 = \mathbb{Q}$. (Exercise 9.) Thus, once one has constructed all primitive pth roots of unity for all primes p less than a given prime P one has all the necessary tools for constructing a primitive Pth root—and hence all primitive Pth roots because the others are the powers of any given one.

§26   The proof sketched above that the pth roots of unity for any prime p can be expressed algebraically in terms of whole numbers, the four arithmetic operations, and the extraction of roots, is entirely constructive. It tells precisely how, with sufficient diligence and patience, one could find the actual expressions. However, in practice these expressions are long, cumbersome, and unilluminating, and are not worth the trouble of finding them. The importance of the constructive proof lies in the information it gives about the *nature* of the formulas, not in the formulas themselves.

An interesting consequence of Gauss's proof is the fact that *the regular 17-gon can be constructed with ruler and compass*. This follows from the observation that if $p = 17$ then $p - 1 = 2 \cdot 2 \cdot 2 \cdot 2$, so that in the sequence of fields $\mathbb{Q} = K_1 \subset K_2 \subset K_4 \subset K_8 \subset K_{16} = \mathbb{Q}(\alpha)$, each extension can be achieved by taking the square root of a known quantity. Since the operation of taking a square root is easy to accomplish geometrically by a ruler and compass construction (see Exercise 14) it follows that the 17-gon can be constructed. The discovery of this fact, which Gauss made when he was only 18 years old, is said to have caused him to decide on a career as a mathematician. In this particular case, $p = 17$, Gauss gave the complete formulas for the roots of unity in the *Disquisitiones Arithmeticae* (Article 354).

In the same way, the regular p-gon can be constructed with ruler and compass whenever $p - 1$ is a power of 2. Only five such primes are known, namely, 3, 5, 17, 257, and 65537. These primes are called *Fermat* primes. (See [E1], pp. 23–25.)

Summary

§27   In Lebesgue's opinion, Vandermonde had all the techniques necessary for the construction of pth roots of unity at his command in the 1770's. Lebesgue even believed that Gauss knew what Vandermonde had done and failed to acknowledge it only because Vandermonde gave no proofs. Such speculations are almost impossible to confirm or refute.* Two questions of the same sort that can probably never be answered are: Was Gauss influenced by Lagrange's *Réflexions*, and, if so, in what way? Gauss's use of Lagrange's letter t for what we call the Lagrange resolvent (Art. 360) suggests a connection. To what extent, if any, was Galois influenced by Gauss's solution of the cyclotomic equation? Galois does mention Gauss's solution of the "binomial equation"—that is, the cyclotomic equation—but only in passing.

The main ideas in the solution of the cyclotomic equation are:

(1) The Lagrange resolvent for the solution of an nth degree equation, that is, the introduction of the quantity $t = x_1 + \beta x_2 + \beta^2 x_3 + \cdots + \beta^{n-1} x_n$, where $x_1, x_2, \ldots, x_n$ are the roots of the equation and β is an nth root of unity.

---

* Note added in second printing: See Loewy, A. "Inwieweit kann Vandermonde als Vorgänger von Gauß bezüglich der algebraischen Auflösung der Kreisteilungsgleichungen $x^n = 1$ angesehen werden?" Jahresber. DMV 27 (1918) 189–195. My thanks to Olaf Neumann for this reference.

(2) Putting the roots of the cyclotomic equation $(x^p - 1)/(x - 1)$ in the order $\alpha, \alpha^g, \alpha^{g^2}, \ldots$ given by a primitive root g mod p. This has the effect of making the $(p - 1)$st power of the Lagrange resolvent of this equation a known quantity (because $\alpha \mapsto \alpha^g$ does not change t^{p-1}, and, by Lemma 2, this means it is independent of α).

(3) Representing all Lagrange resolvents (obtained by using other $(p - 1)$st roots of unity β) in terms of one of them. This is accomplished by noting that $t_i t^{-i}$ is invariant under $\alpha \mapsto \alpha^g$ and is therefore known. ($t \neq 0$ by Lemma 2.) This gives the solution $\alpha = (t + t_2 + t_3 + \cdots + t_{p-1})/(p - 1)$ involving known quantities and the $(p - 1)$st root of a known quantity.

(4) Breaking the extension into extensions of prime order. Instead of a $(p - 1)$st root of unity β in the Lagrange resolvent, one then needs a qth root of unity where q is a prime $< p$. Therefore the prime roots of unity can be found in succession.

None of these ideas seems to have been original with Gauss, but he was the first to put them all together in a complete, clear, and rigorous way. Even he, however, did not—because of the lack of a proof of what is called Lemma 2 above—give a complete proof that pth roots of unity can be expressed in terms of radicals. (The proof of the solvability of the equation by radicals was not, after all, his objective. Moreover, he most probably did have a proof of Lemma 2, even though he never published one.)

Third Exercise Set

1. Give the ruler and compass construction of the cube roots of 1.

2. Show that in order to construct a 5th root of unity (or an nth root of unity for any n) it suffices to find its *real part*. Give a ruler and compass construction of the real part of one of the 5th roots of unity given in §21. Compare to Euclid's construction of the regular pentagon. (Euclid's *Elements*, Book 4, Proposition 11. See also Aaboe [A1], Chapter 2.)

3. Derive the formulas for the 5th roots of unity given in §21. Use them (preferably with a pocket calculator) to calculate $\sin 72°$ and $\cos 72°$ and check the result.

4. Show that if j and k are relatively prime then the product of a primitive jth root of unity and a primitive kth root of unity is a primitive jkth root of unity.

5. Show that the formula $\alpha = \frac{1}{10}[t_1 + (t_2 t_1^8)/t_1^8 + (t_3 t_1^7)/t_1^7 + \cdots]$ of §22, where $t_2 t_1^8, t_3 t_1^7, \ldots$ are known quantities, gives an 11th root of unity when t_1 is replaced by any 10th root of t_1^{10}.

6. Show that if $t = \alpha + \beta\alpha^g + \beta^2\alpha^{gg} + \ldots$ as in §24 then $(p - 1)^{-1}[s + (t_2 t^{p-3})/s^{p-3} + (t_3 t^{p-4})/s^{p-4} + \ldots$ is a primitive pth root of unity for any solution s of $s^{p-1} = t^{p-1}$.

7. Prove Lemma 1 of §24, which states that every prime has a primitive root.

8. Prove that if $a_0, a_1, a_2, \ldots, a_n$ are rational numbers, if α is a pth root of unity $\alpha \neq 1$, and if $a_0 + a_1\alpha + a_2\alpha^2 + \cdots + a_n\alpha^n = a_0 + a_1\alpha^g + a_2\alpha^{2g} + \cdots + a_n\alpha^{ng}$, then $a_0 + a_1\alpha + a_2\alpha^2 + \cdots + a_n\alpha^n$ is equal to a rational number. Use the fact that α is not the root of any nonzero polynomial of degree $< p - 1$ with integer coefficients. (This fact will be proved in §64 and in Exercise 8 of the eighth set.)

9. In the case $p = 19$ of the construction of §25, find a sequence of two intermediate fields $K_1 \subset K_d \subset K_{d'} \subset K_{18}$ where the inclusions are all strict inclusions. Show that for any p the maximum number of such intermediate fields is equal to one less than the number of prime factors of $p - 1$, counted with multiplicities.

10. Use the method of §25 to derive the formulas of §21 for $\sqrt[5]{1}$. (In the first stage set $\gamma_1 = \alpha$ and $d = 2$ to find a resolvent t_1 whose square is invariant under S^2 where $S: \alpha \rightarrow \alpha^2$. In the second stage set $\gamma_2 = \alpha + \alpha^4$, and $d = 2$ to find a resolvent t_2 whose square is rational.)

11. Gauss's method of §25 applied to the case $p = 7$, $D = 6$, $d = 2$ gives an algebraic expression for the solution $\gamma = \alpha + \alpha^{-1}$ of the equation $\gamma^3 + \gamma^2 - 2\gamma - 1 = 0$ of the beginning of §22. Find this expression (which involves the cube root of unity ω).

12. Derive Gauss's algebraic expression for $\cos(2\pi/17)$ in Section 365 of *Disquisitiones Arithmeticae*:

$$\cos(2\pi/17) = -\frac{1}{16} + \frac{1}{16}\sqrt{17} + \frac{1}{16}\sqrt{34 - 2\sqrt{17}}$$

$$+ \frac{1}{8}\sqrt{17 + 3\sqrt{17} - \sqrt{34 - 2\sqrt{17}} - 2\sqrt{34 + 2\sqrt{17}}}.$$

[Evaluate numerically the elements of the fields K_1, K_2, K_4 in succession and use this to evaluate the needed cosine $(\alpha + \alpha^{-1})/2$, which is in K_8. The main technique is the resolvent method described in §25. However, the problem is not entirely algebraic because the *signs* of the radicals in Gauss's formula must be determined by nonalgebraic means.]

13. Give algebraic expressions of all eight values of $\cos(2\pi k/17)$ other than the trivial value 1.

14. Find a ruler and compass construction for the *square root*, or, as Euclid would express it, given a rectangle, construct a square with the same area. (Euclid's *Elements*, Book II, Proposition 14.)

15. Show that if α is a primitive pth root of unity (p = prime) then the p^{n-1}st root of α is a primitive p^nth root of unity.

Galois Resolvents

§28   For Lagrange, the "resolvent" of an equation had three crucial properties.

(1) It is rationally expressible in terms of the roots of the equation and known quantities (including rational numbers, the coefficients of the given equation, and roots of unity).

(2) Conversely, each of the roots of the equation can be expressed rationally
in terms of it and known quantities.

(3) It is the solution of a *solvable* equation.

Here (1) describes the universe from which the resolvent is to be selected,
and (2) and (3) describe the properties of the resolvent through which it
solves the equation.

Lagrange doubted that it was possible to find a resolvent for the general
5th degree equation. However, he did not leave it at that. He gave serious
consideration to the question of determining which quantities satisfying (1)
also satisfied (2), and he proved in this connection a very important theorem
which, as will be shown below, can be taken as the basis of Galois theory. In
fact, Galois' idea can be briefly described by saying that he realized that a
"resolvent" with properties (1) and (2) exists in all cases, and this "resolvent"
can be used to describe the *form* of the solutions fully enough to show that,
as Lagrange suspected, no resolvent with all three properties exists.

§29   Lagrange states the theorem in question as follows in his *Réflexions*
(Article 104):

"*If t and y are any two functions* [*polynomials*] *in the roots* x', x'', x''', ...
of $x^\mu + mx^{\mu-1} + nx^{\mu-2} + [px^{\mu-3} +] \cdots = 0$ *and if these functions are such
that every permutation of the roots* x', x'', x''', ... *which changes y also changes
t, one can, generally speaking, express y rationally in terms of t and m, n,
p,* ..., *so that when one knows a value of t one will also know immediately the
corresponding value of y; we say* generally speaking *because if the known value
of t is a double or triple or higher root of the equation for t then the corresponding
value of y will depend on an equation of degree 2 or 3 or higher with coefficients
that are rational in t and m, n, p,* .... "

This theorem is Lagrange's answer—and it is a very satisfactory answer—
to the problem of determining which elements with property (1) of the preceding article also have property (2). Rather than asking whether all *roots* can
be expressed rationally in terms of *t* and known quantities, Lagrange asks
whether a *function* *y* of the roots (which could be a root) can be expressed
rationally in terms of *t*. If *y* can be expressed in terms of *t*, then surely any
permutation of the roots which leaves *t* unchanged will not change *y*. The
theorem says that this necessary condition is, "generally speaking", sufficient,
that is, that if every permutation of the roots which leaves *t* unchanged also
leaves *y* unchanged then *y* can be expressed in terms of *t*.

In order to see the situation in which the "generally speaking" provision
is necessary, let *t* be a polynomial in the roots x', x'', x''', ... of the given equation with coefficients that are known quantities. If ϕ is any permutation of
x', x'', x''', ... let ϕt be the polynomial in x', x'', x''', ... obtained by applying
ϕ to the *x*'s in *t*. It may be that $\phi t = t$ even when $\phi \neq$ identity. (For example,

when $n = 4$ and $t = x' + x'' - x''' - x^{(iv)}$ as in §17, the twenty-four permutations of the x's produce only six different polynomials ϕt.) The "equation for t" is the polynomial $F(X) = (X - t)(X - \phi_1 t)(X - \phi_2 t) \cdots (X - \phi_k t)$, where $t, \phi_1 t, \phi_2 t, \ldots$ is a list of all *distinct* polynomials that can be obtained from t by permutations of the x's. The coefficients of $F(X)$ are symmetric polynomials in $t, \phi_1 t, \phi_2 t, \ldots$, and, since a permutation of $x', x'', x''', \ldots$ merely permutes $t, \phi_1 t, \phi_2 t, \ldots$, they are therefore *symmetric* polynomials in $x', x'', x''', \ldots$, which means that *the coefficients of $F(X)$ are known quantities.* Thus the polynomial $F(X)$ is a known polynomial of which t is a root. Lagrange says that if t is a *simple* root of $F(X)$ and if $\phi t = t$ implies $\phi y = y$ then y can be expressed rationally in terms of t and known quantities. If t is a *multiple* root of $F(X)$ (since $F(X) = \prod(X - \phi_i t)$ this means that one of the polynomials $\phi_i t$ has, although it is *formally* distinct from t, the same *numerical* value when numerical values of the roots $x', x'', x''', \ldots$ are used) then y is not a rational function of t, but is a root of a polynomial whose coefficients are rational functions of t and whose degree is equal to the multiplicity of t as a root of $F(X)$.

For the sake of simplicity, Lagrange's theorem will be proved only in the case where $F(X)$ has *only* simple roots, that is, where the formally distinct polynomials $t, \phi_1 t, \phi_2 t, \ldots$ are all numerically distinct. This is the only case that is necessary for Galois theory.

§30   PROOF OF LAGRANGE'S THEOREM IN THE CASE WHERE THE EQUATION FOR t HAS SIMPLE ROOTS. Let t and y be given polynomials in the n roots, and let $t_1, t_2, \ldots, t_k$ be all the distinct polynomials in the roots that can be obtained from t by permutation of the roots. By assumption, any permutation which leaves t unchanged leaves y unchanged. Therefore, there are *at most* k different polynomials that can be obtained from y by permuting the variables, and there are polynomials $y_1, y_2, \ldots, y_k$ such that the permutations which carry t to t_i carry y to y_i. (Explicitly, let $\phi_1, \phi_2, \ldots, \phi_k$ be permutations such that $t_i = \phi_i t$ and set $y_i = \phi_i y$. Every permutation ψ can be written in the form $\psi = \phi_i g$ where $i = 1$ or 2 or ... or k and where g is a permutation which leaves t unchanged (see Exercise 1). Then $gy = y$ and $\psi y = \phi_i y = y_i$, so the list $y_1, y_2, \ldots$, includes all polynomials of the form ψy.)

The key to Lagrange's theorem is to consider the k polynomials

$$y_1 + y_2 + \cdots + y_k,$$

$$t_1 y_1 + t_2 y_2 + \cdots + t_k y_k,$$

$$t_1^2 y_1 + t_2^2 y_2 + \cdots + t_k^2 y_k, \tag{1}$$

$$\vdots$$

$$t_1^{k-1} y_1 + t_2^{k-1} y_2 + \cdots + t_k^{k-1} y_k.$$

These polynomials are symmetric (a permutation of $x', x'', x''', \ldots$ merely permutes the terms of each sum) and therefore their numerical values are known (expressible in terms of the coefficients of t and y and in terms of the coefficients $m, n, p, \ldots$ of the given equation). What is to be shown is that y_1 can be expressed rationally in terms of t_1 and known quantities. For this, it suffices to apply Cramer's rule. (Lagrange's approach is more inclusive because he does not make the simplifying assumption that $F(X)$ has simple roots.)

Cramer's rule applied to the equations $\sum t_i^j y_i = c_j$, where c_j is a known quantity, gives $y_\mu = D_\mu/\Delta$ where Δ is the Vandermonde determinant $\det(t_i^j) = \prod_{k<i}(t_i - t_k)$ and where D_μ is the determinant obtained by replacing the μth column of Δ by the column of known quantities c_j. The square of the denominator $\Delta^2 = \prod_{k<i}(t_i - t_k)^2$ is simply the discriminant of $F(X) = \prod(X - t_i)$; by assumption this is not 0, and of course it is a known quantity. It suffices, then, to show that the numerator of $y_1 = D_1\Delta/\Delta^2$ can be expressed rationally in terms of t_1. In fact, it is easy to see that $D_1\Delta$ is a polynomial in t_1. On its face, it is a polynomial in all the t_i (and in the known quantities c_i). However, permutation of two of the t_i other than t_1 changes the sign of both D_1 and Δ, and therefore leaves $D_1\Delta$ unchanged. Thus $D_1\Delta$ can be expressed as a polynomial in t_1 whose coefficients are symmetric polynomials in $t_2, t_3, \ldots, t_k$. These coefficients can, by the fundamental theorem on symmetric functions, be expressed in terms of the elementary symmetric functions in $k-1$ variables $\tau_1 = t_2 + t_3 + \cdots$, $\tau_2 = t_2 t_3 + t_2 t_4 + \cdots$, $\ldots, \tau_{k-1} = t_2 t_3 \ldots t_k$. But, as was seen in §11, $\tau_1 = \sigma_1 - t_1, \tau_2 = \sigma_2 - t_1\sigma_1 + t_1^2$, $\tau_3 = \sigma_3 - t_1\sigma_2 + t_1^2\sigma_1 - t_1^3, \ldots$, where the σ's, being symmetric functions of the t's, are known quantities. Thus the τ's and consequently $D_1\Delta$ can be expressed in terms of t_1 and known quantities, as was to be shown.   □

§31 Corollary. *If one can find a polynomial t in the roots $x', x'', x''', \ldots$ with the property that its $n!$ different permutations are not merely* formally *distinct (this would be easy to accomplish—for example by setting $t = x' + 2x'' + 3x''' + \cdots$) but also* numerically *distinct, then every permutation of the roots changes t and the condition of Lagrange's theorem is met by* all *polynomials y. Therefore all polynomials in the roots, and, in particular, the roots themselves can be expressed rationally in terms of (any) one numerical value of t.*

In short, if such a t can be found, then it is a "resolvent" in the sense that (1) and (2) of §28 are satisfied. Such a t is called a "Galois resolvent" because Galois was the first* to realize that there always exist such t's and to realize how useful they are in studying the solution of an equation even when they do not satisfy (3) of §28. Of course it is impossible to find such a t if two or more of the roots $x', x'', \ldots$ coincide, so Galois must—and does—exclude equations in which this occurs. (This is not a real restriction because if $f(x) = 0$ has multiple roots then it has roots in common with $f'(x) = 0$ and

---

* In his article [K1] on the history of Galois theory, Kiernan erroneously states (p. 81) that Lagrange had proved the existence of a Galois resolvent.

the greatest common divisor of f and f', which is a polynomial that can be found using the Euclidean algorithm (§35), has as its roots the multiple roots of f; then f divided by this greatest common divisor has the same roots as f but they all occur with multiplicity 1. See Exercise 2.)

Galois does not *prove* the existence of a Galois resolvent, he merely asserts it (see Appendix 1). Moreover, he asserts that there exist Galois resolvents of the special form $t = A'x' + A''x'' + A'''x''' + \cdots$, where the coefficients $A', A'', A''', \ldots$ are suitably chosen *integers*. This is certainly a plausible statement. Equality of two permutations of $A'x' + A''x'' + A'''x''' + \cdots$ is an exceptional circumstance, one that imposes a condition on the A's. One need only choose the A's in such a way that none of these conditions is satisfied. The proof which follows is a carrying out of this idea. Since Galois gives no indication at all of a proof, there is no way of knowing whether this proof resembles what he had in mind.

§32 Henceforth, let the n roots of the given equation be denoted a, b, c, ... The assumption is that these roots are *distinct*, which is to say that the discriminant $(a - b)^2(a - c)^2(b - c)^2 \cdots = D$ is not zero. (See §13.) What is to be shown is that n integers $A, B, C, \ldots$ can be chosen so that the $n!$ numbers $AS(a) + BS(b) + CS(c) + \cdots$ are distinct, where S ranges over all $n!$ permutations of the roots $a, b, c, \ldots$. Let $\mathscr{D}$ be the product of the squares of the differences of these $n!$ numbers, that is,

$$\mathscr{D} = \prod_{S, T} [A(S(a) - T(a)) + B(S(b) - T(b)) + \cdots]^2,$$

where the product is over all $n!\,(n! - 1)/2$ pairs (unordered) of permutations S and T in which $S \neq T$. Let $A, B, C, \ldots$ be regarded at first as variables. Then $\mathscr{D}$ is a polynomial in $A, B, C, \ldots$ whose coefficients are polynomials in the roots $a, b, c, \ldots$. Since $\mathscr{D}$ is obviously symmetric in the roots, these coefficients are symmetric in the roots, so $\mathscr{D}$ is a polynomial in $A, B, C, \ldots$ with known coefficients. Since $\mathscr{D}$ is a product of nonzero polynomials, it is nonzero (Exercise 4). Therefore (Exercise 3) one can assign integer values to $A, B, C, \ldots$ in such a way as to make $\mathscr{D} \neq 0$, as was to be shown.

§33 The basic use of the Galois resolvent is the following.

Let K denote* the set of all quantities that are to be regarded as known. K must include at least the rational numbers and the coefficients of the given equation $f(x) = 0$, but it may include more—for example, there may be some reason to regard certain roots of unity as known. Since sums, differences, products, and quotients (with nonzero denominator) of known quantities are known, K is what is called a "field." (A field is a set with two operations— addition and multiplication—in which the familiar associative, commutative,

---

* The fact that K is the initial letter of "known" is a pleasant coincidence. The use of K comes from the fact that the set is what is known in German as a "Körper," and in English as a "field."

and distributive laws hold, and in which subtraction and division—determination of x by $a + x = b$ or by $cx = d$ when $c \neq 0$—are possible.)

Let $K(a, b, c, \ldots)$ denote* the set of all elements which can be expressed as polynomials in the roots a, b, c, ... of the given equation $f(x) = 0$ with coefficients in K. (The precise domain in which a, b, c, ..., and hence the elements of $K(a, b, c, \ldots)$, have their existence will be described later. At first, it is probably easiest to think of $f(x)$ as having *rational* coefficients, of K as being the rational numbers, and of $a, b, c, \ldots$ as being complex numbers. Ultimately, however, it will be important to be able to think of the coefficients of $f(x)$ as being literal, that is, as being elements of K that are transcendental over the rationals. All that matters is that there be some field $K(a, b, c, \ldots)$ containing K and containing a complete set of roots a, b, c, ... of $f(x) = 0$. Such a field is called a *splitting field* of f.)

In particular, t is in $K(a, b, c, \ldots)$. Let $K(t)$ denote the set of all elements of $K(a, b, c, \ldots)$ which can be expressed rationally in terms of t with coefficients in K. Then the corollary of §31 can be expressed by the equation

$$K(a, b, c, \ldots) = K(t),$$

that is, everything that can be expressed as a polynomial in the roots a, b, c, ... can be expressed rationally in terms of t and *vice versa*. (It is not obvious that a rational function of t can necessarily be expressed as a polynomial in a, b, c, ..., only that it can be expressed as a rational function in them. However, as is noted at the end of §36, every rational function in a, b, c, ... can be expressed as a polynomial in a, b, c, ...) This can be regarded as a *solution* of the equation $f(x) = 0$ because, as will be shown in the next three sections, $K(t)$ can be described in a very explicit way, and it is a *field*. Thus, the construction of $K(t)$ which follows is a construction of a field $K(t) = K(a, b, c, \ldots)$ which contains roots a, b, c, ... of the given equation.

(Strictly speaking, the construction of $K(t)$ which follows *assumes* that there is some field in which $f(x) = 0$ has n roots, so that, for example, $K(a, b, c, \ldots)$ makes sense and provides a setting in which the computation of the polynomial $F(X) = \prod_S (X - St)$ can be carried out. In other words, it *constructs* a field containing roots assuming *there exists* a field containing roots. This assumption, which Girard, Newton, Lagrange, Galois, *et al.* seem to have made without question, will be justified in §§49–61.)

Construction of the Field $K(t)$

§34   Let $f(x) = 0$ be an equation of degree n with n distinct roots a, b, c, ... and let t be a Galois resolvent of this equation. Then $K(a, b, c, \ldots) = K(t)$. The objective is to describe the field $K(t)$ as explicitly as possible. By

---

* This notation should not be confused with the notation for functions. Here K is a *set* (the known quantities) and $K(a, b, c, \cdot \cdot \cdot)$ is a *larger set*. Similarly, in the next paragraph, $K(t)$ is a set containing both K and the element t, and the notation has nothing to do with functions.

assumption, t is a polynomial in $a, b, c, \ldots$ which has the property that the $n!$ elements St of $K(a, b, c, \ldots)$ obtained by substituting the roots $a, b, c, \ldots$ into t in each of their $n!$ possible orders are *distinct* elements of $K(a, b, c, \ldots)$. (One can, in fact, assume that t is a linear polynomial with integer coefficients.) The "equation for t," as in §29, is the polynomial* $F(X) = \prod_S (X - St)$, where the product is over all $n!$ permutations S of the roots. The coefficients of $F(X)$ are, in the first instance, in $K(a, b, c, \ldots)$, but since they are symmetric polynomials in $a, b, c, \ldots$ they can, by the fundamental theorem on symmetric polynomials, be expressed in terms of the elementary symmetric polynomials in the roots $a, b, c, \ldots$ and therefore in terms of the coefficients of the given equation $f(x) = 0$. Thus, $F(X)$ *is a polynomial with known coefficients in K.*

(The computation of $F(X)$ is stupendously long in all but the simplest cases, but in principle it can be carried out. Given the integers $A, B, C, \ldots$ which describe $t = Aa + Bb + Cc + \cdots$, the coefficients of $F(X)$ are explicit symmetric polynomials in $a, b, c, \ldots$ with integer coefficients. Express them as polynomials in the elementary symmetric polynomials $\sigma_1, \sigma_2, \ldots, \sigma_n$ in $a, b, c, \ldots$. If the given equation is $B_0 x^n + B_1 x^{n-1} + \cdots + B_n = 0$ then $\sigma_i = (-1)^i B_i/B_0$ for $i = 1, 2, \ldots, n$, that is,

$$f(x) = B_0(x - a)(x - b)(x - c) \ldots.$$

This describes each σ_i as an element of K, and substitution of these values in the symmetric polynomials gives the coefficients of $F(X)$ as elements of K.)

The polynomial $F(X)$ with coefficients in K has degree $n!$ and leading coefficient 1. Let $F(X) = G_1(X)G_2(X) \ldots G_k(X)$ be its factorization into *irreducible* polynomials $G_1(X), G_2(X), \ldots, G_k(X)$ with coefficients in K and leading coefficient 1. (A polynomial $G(X)$ with coefficients in K is said to be *irreducible* if there do not exist two polynomials $g(X), h(X)$ with coefficients in K such that $G(X) = g(X)h(X)$ and such that g and h are of lower degree than G. If one accepts the "obvious fact" that a polynomial is either irreducible or there is a factorization of it into two polynomials of lower degree then it is obvious that such a factorization of F always exists.† For a full discussion of the problem of factorization see §§49–61.) The substitution of t for X in $F(X)$ gives,‡ on the one hand, $F(t) = 0$ because $F(t)$ can be

---

* In the remainder of the book, x is used to denote the variable in the original equation $f(x) = 0$, and X is used to denote the variable in auxiliary equations relating to the Galois resolvent. Of course there is no logical necessity to use different letters in the two cases, but it seems clearer to do so.

† The catch is that one might be unable to factor the given polynomial and at the same time be unable to prove that no factorization exists.

‡ Here t is regarded as an element of $K(a, b, c, \ldots)$. Strictly speaking, the same letter t should not be used to represent both the polynomial in variables and the element of $K(a, b, c, \ldots)$ obtained by replacing the variables by the roots $a, b, c, \ldots$ of the given equation. In what follows, t for the most part will represent the element of $K(a, b, c, \ldots)$. It is only in the expression St, which represents the value of the polynomial when the roots $a, b, c, \ldots$ are substituted in the order indicated by S, that t must be considered to be a polynomial in variables. (See Appendix 2, where T is used to denote the polynomial and $t_1, t_2, \ldots, t_{n!}$ denote its $n!$ images in $K(a, b, c, \ldots)$.)

written as a product $\prod_S (t - St)$ in which one of the factors is 0, and gives, on the other hand, $F(t) = G_1(t)G_2(t) \ldots G_k(t)$. Therefore $G_1(t)G_2(t) \ldots G_k(t) = 0$ and at least one of the factors $G_i(t)$ must be* 0. In other words, the irreducible factors $G_1, G_2, \ldots, G_k$ of F must include an irreducible polynomial of which t is a root. The desired description of $K(t)$ is then given by the following basic theorem:

Theorem (Simple Algebraic Extensions). *Let K be a given field and let G(X) be an irreducible polynomial with coefficients in K. Then one can construct a field K(t) such that*

(1) *$K(t)$ contains K,*
(2) *$K(t)$ contains an element t which is a root of G, that is, which satisfies $G(t) = 0$, and*
(3) *every element of $K(t)$ can be expressed as a polynomial in t with coefficients in K, that is, given any $x \in K(t)$ there are an integer v and elements b_0, $b_1, \ldots, b_v$ of K such that $x = b_0 + b_1 t + \cdots + b_v t^v$.*

Moreover, any two such fields $K(t)$ are the *same* in a natural sense that will be spelled out in §36.

Roughly speaking, the field $K(t)$ to be constructed can be described as follows. Its *elements* are polynomials of degree $< \deg G$ in t with coefficients in K. Elements are added in the natural way. Elements are multiplied by multiplying them as polynomials and then using the relation $G(t) = 0$ to reduce the product to a polynomial of degree $< \deg G$. (For example, if $G(t) = t^2 + 1$ then the product of $at + b$ and $ct + d$ is $act^2 + adt + bct + bd = ac(t^2 + 1) - ac + adt + bct + bd = (ad + bc)t + (bd - ac)$. This is, of course, the rule for multiplying complex numbers—here $t = \sqrt{-1}$.) Clearly the set $K(t)$ with the operations of addition and multiplication defined in this way satisfies (1)—because the polynomials of degree 0 correspond one-to-one to elements of K—and (2)—because $G(t) = 0$ by the definition of $t^{\deg G}$—and (3), by the very definition of the elements. The problem is to show that $K(t)$ defined in this way is a *field*, and the hard part here is to show that *division by nonzero elements is possible*. The basis of the proof that $K(t)$ is a field is the *Euclidean algorithm*. This algorithm, which is one of the most useful tools in algebra and number theory, is the subject of the next article.

§35   Let $a(X)$ and $b(X)$ be polynomials with coefficients in a field K. The Euclidean algorithm is a method of finding a common divisor $d(X)$ of $a(X)$ and $b(X)$ which can be expressed in the form $d(X) = A(X)a(X) + B(X)b(X)$,

---

* This "obvious fact" follows, in the last analysis, from the fact that the splitting field $K(a, b, c, \ldots)$ is a field and therefore has no zero divisors. For now, let the elements of $K(a, b, c, \ldots)$, i.e. polynomials in $a, b, c, \ldots$ with coefficients in K, be regarded as *numbers*. Then the needed conclusion follows from the fact that a product of two numbers is zero only if one of the factors is zero.

where $A(X)$ and $B(X)$, and consequently $d(X)$, are polynomials with co-efficients in K. Then $d(X)$ is clearly a *greatest common divisor* of $a(X)$ and $b(X)$ in the sense that any polynomial with coefficients in K which divides both $a(X)$ and $b(X)$ also divides $d(X) = A(X)a(X) + B(X)b(X)$.

The Euclidean algorithm is based on the algorithm for *division with remainder*. This is an algorithm which enables one to find, given two poly-nomials $a(X)$ and $b(X) \neq 0$ with coefficients in K, two other polynomials, the *quotient* $q(X)$ and the *remainder* $r(X)$, also with coefficients in K, such that $a(X) = q(X)b(X) + r(X)$ and such that either $r(X) = 0$ or the degree of $r(X)$ is less than the degree of $b(X)$. (In this context it is convenient to adopt the convention that the polynomial 0 has degree $-\infty$ so that one can simply say deg $r <$ deg b.) The existence and uniqueness of $q(X)$ and $r(X)$ are easy to prove by induction on the degree of $a(X)$ since it is easy to see that if deg $a >$ deg b then subtraction of a suitable multiple of b from a will reduce its degree. (See Exercise 18.) An algorithm for division of polynomials that is analogous to ordinary long division—called *synthetic division*—is easy to derive (Exercise 19).

In terms of the division algorithm, the Euclidean algorithm can be described as follows. Let $a(X)$ and $b(X)$ be given and assume without loss of generality that deg $a \geq$ deg b. Use the division algorithm to write $a = q_1 b + r_1$. If $r_1 \neq 0$, use the division algorithm to write $b = q_2 r_1 + r_2$. If $r_2 \neq 0$, use the division algorithm to write $r_1 = q_3 r_2 + r_3$, and so forth. Since deg $b >$ deg $r_1 >$ deg $r_2 > \cdots$, eventually the remainder of the division that is called for by the algorithm is 0. Let $d(X)$ be the divisor of the division that results in the remainder 0. Thus, if $r_1 = 0$, then $d(X) = b(X)$, and otherwise $d(X) = r_i(X)$ where $r_i(X)$ is the last nonzero remainder. It is to be shown that $d(X)$ divides both $a(X)$ and $b(X)$ and that it can be written in the form $d(X) = A(X)a(X) + B(X)b(X)$.

Let $r_0(X)$ denote $b(X)$ and $r_{-1}(X)$ denote $a(X)$. Then the divisions of the Euclidean algorithm are

$$r_{-1} = q_1 r_0 + r_1,$$

$$r_0 = q_2 r_1 + r_2,$$

$$r_1 = q_3 r_2 + r_3,$$

$$\cdots$$

$$r_{i-2} = q_i r_{i-1} + r_i,$$

$$r_{i-1} = q_{i+1} r_i,$$

where $r_i = d$. The last equation shows that d divides r_{i-1}. Then the next-to-last equation shows that d divides r_{i-2} (because it divides both terms on the right), the third from last shows that d divides r_{i-3}, and so forth. Thus d divides all r's, including $r_0 = b$ and $r_{-1} = a$, as was to be shown.

When the first equation is written in the form $r_1 = r_{-1} - q_1 r_0 = a - q_1 b$, it shows that r_1 can be written as a linear combination of a and b. When the second equation is written in the form $r_2 = r_0 - q_2 r_1$ and the expressions $r_0 = b$ and $r_1 = a - q_1 b$ of r_0 and r_1 in terms of a and b are substituted into it, it gives an expression of r_2 in terms of a and b. In the same way, the jth equation $r_j = r_{j-2} - q_j r_{j-1}$ shows that a linear expression of r_j in terms of a and b can be deduced from similar expressions of r_{j-2} and r_{j-1}. Thus $d = r_i$ can be expressed linearly in terms of a and b, $d = Aa + Bb$, as required.

§36   PROOF OF THE THEOREM ON SIMPLE ALGEBRAIC EXTENSIONS (§34). As in the statement of the theorem, let K be a given field and let $G(X)$ be an irreducible polynomial with coefficients in K. Let R be the set of all polynomials in the variable X with coefficients in K. The elements of R can be added and multiplied in the obvious way. Two elements of R will be said to be *equivalent* (congruent mod G) if their difference is divisible by G, that is, if their difference can be written as an element of R times $G \in R$. This defines an equivalence relation on R (the relation is reflexive, symmetric, and transitive). Let L denote the set of equivalence classes. Then, because the equivalence relation is consistent with both addition and multiplication (if $a(X)$ is equivalent to $b(X)$ then $a(X) + c(X)$ is equivalent to $b(X) + c(X)$ and $a(X)c(X)$ is equivalent to $b(X)c(X)$ for all $c(X)$), elements of L can be added and multiplied. The mapping $K \to L$ which carries $k \in K$ to the equivalence class containing the polynomial k of degree 0 is a one-to-one map, by means of which one can regard K as being *contained* in L. Let t denote the class of the polynomial X. Then, by definition, $G(t)$ is the class of $G(X)$, which is the class of 0, which is identified with $0 \in K$. Finally, if y is any element of L then y is the class of some polynomial $h(X) = B_0 X^m + B_1 X^{m-1} + \cdots + B_m$; then $h(t) = B_0 t^m + B_1 t^{m-1} + \cdots + B_m$ is the class of $B_0 X^m + \cdots + B_m = h(X)$, that is, $h(t) = y$. Therefore L satisfies conditions (1), (2) and (3) of the theorem. It is to be shown that L *is a field*.

Elements of L can be added and multiplied in natural ways and all the axioms of arithmetic are obviously satisfied with the possible exception of the axiom that *division by nonzero elements is possible*. Let y be a nonzero element of L, say the class of $h(X)$. By the Euclidean algorithm, one can find a polynomial $d(X)$ which divides both $h(X)$ and $G(X)$ and which can be expressed in the form $d(X) = A(X)h(X) + B(X)G(X)$, where $A(X)$ and $B(X)$ are polynomials with coefficients in K. Since $d(X)$ divides $G(X)$ and $G(X)$ is irreducible, $d(X)$ is either a nonzero constant in K or it is $G(X)$ times a nonzero constant in K. If it were $G(X)$ times a nonzero constant of K then, since it divides $h(X)$, $h(X)$ would be equivalent to 0 and y would be 0. Since y is nonzero by assumption, it follows that $d(X)$ is a nonzero constant, call it $d \in K$. From $d = A(X)h(X) + B(X)G(X)$ it follows that $1 = d^{-1}A(X)h(X) + d^{-1}B(X)G(X)$. Thus the element 1 of L, which is to say the class of the polynomial 1, is the same as the class of $d^{-1}A(X)h(X)$, which is the class of $d^{-1}A(X)$ times y. In short, the class of $d^{-1}A(X)$ is a multiplicative inverse of y, and division by y is possible, as was to be shown.

It remains only to specify the sense in which L is unique. Let L' be any field satisfying (1), (2), and (3) for some $t' \in L'$ which is a root of G. The claim is that then *there is a natural isomorphism $L \to L'$*, that is, a function $L \to L'$ which preserves addition and multiplication and which is one-to-one and onto. For this, let R be mapped to L' by sending $h(X) \in R$ to the element of L' obtained by substituting t' for X and computing $h(t')$ in L'. This mapping preserves addition and multiplication. It is onto by property (3) of L'. If $h_1(X)$ and $h_2(X)$ are equivalent, say $h_1(X) - h_2(X) = q(X)G(X)$, then $h_1(t') - h_2(t') = q(t')G(t') = 0$. Thus the function $R \to L'$ is compatible with equivalence, and defines a function $L \to L'$. It remains only to show that this function is one-to-one. If the classes y_1 and y_2 have the same image in L' then the class $y_1 - y_2$ has the image 0. It is to be shown that this happpens only when $y_1 - y_2 = 0$. This amounts to saying that any polynomial $h(X)$ for which $h(t') = 0$ must be divisible by $G(X)$. Let the Euclidean algorithm be used to find a common divisor $d(X) = A(X)h(X) + B(X)G(X)$ of h and G. If $h(t') = 0$ then $d(t') = A(t') \cdot 0 + B(t') \cdot 0 = 0$. Therefore $d(X)$, which is nonzero by definition, cannot be a polynomial of degree 0. Since $d(X)$ divides $G(X)$ and $G(X)$ is irreducible, this implies that $d(X)$ is a nonzero multiple of $G(X)$. Since $d(X)$ divides $h(X)$, it follows that $G(X)$ divides $h(X)$, as was to be shown. □

In summary, it has been shown how to find, for a given equation $f(x) = 0$ with coefficients in K, a Galois resolvent t and an irreducible polynomial $G(X)$ with coefficients in K of which t is a root. Here it is assumed that there is some large field containing K in which $f(x) = 0$ has $\deg f$ roots $a, b, c, \ldots$, and that these roots are distinct. By the Corollary of §31, all the roots can be expressed as rational functions of t, that is, the roots $a, b, c, \ldots$ can be expressed as elements of the field $K(t)$ obtained by adjoining a root t of the irreducible polynomial $G(X)$ to K using the theorem of §34. Thus $K(t)$ contains $\deg f$ roots of $f(X)$, and the construction of the field $K(t)$ represents in this sense a *solution* of f.

Of course, this is not at all the sort of solution one had in mind at the outset, and in particular it leaves entirely unanswered the question of generalizing to higher degrees the solution by radicals of quadratic, cubic, and biquadratic equations. What Galois did was to show how a solution by a Galois resolvent can be made to yield enough information to answer many questions about the algebraic nature of the solution and, in particular, the question of whether a solution can be expressed in terms of radicals.

Note that $K(t)$ is a field and that every element of $K(t)$ can be expressed as a polynomial in t with coefficients in K. Therefore, not only polynomials in the roots a, b, c, ... with coefficients in K (the elements of the set $K(a, b, c, \ldots)$ of §33), but in fact *rational functions* of the roots $a, b, c, \ldots$ with coefficients in K can (provided the denominator is nonzero) be expressed as elements of $K(t)$, that is, as *polynomials* in t with coefficients in K (and therefore as polynomials in a, b, c, ...).

Galois' Proof of the Basic Lemma

§37   The main tool in the above expression of $K(a, b, c, \ldots)$ in the simple form $K(t)$ is Lagrange's theorem of §29, which is used to show that the roots $a, b, c, \ldots$ can all be expressed as polynomials in the Galois resolvent t. This, however, is not the approach that Galois himself took. He mentioned neither Lagrange nor Lagrange's theorem and instead stated and proved the crucial fact as follows:

"**Lemma III.** *When the function* t has been chosen as above* [*that is, so that it has n! different values when the roots are permuted*], *it has the property that all the roots of the equation can be expressed rationally in terms of t.*

"In fact, let $t = \phi(a, b, c, \ldots)$, which is to say $t - \phi(a, b, c, \ldots) = 0$. Let us multiply all of the similar equations [sic] which one obtains in permuting the roots, keeping only the first one fixed. This gives

$$(t - \phi(a, b, c, d, \ldots))(t - \phi(a, c, b, d, \ldots))(t - \phi(a, b, d, c, \ldots)) \cdots,$$

which is symmetric in $b, c, d, \ldots$, and consequently can be written as a function of a. We will then have an equation of the form $F(t, a) = 0$. But I say that one can solve this for a. For this it suffices to seek the common solution of this equation and the given one. This solution [that is, a] is the only one they have in common, because one cannot have, for example, $F(t, b) = 0$ without the consequence (this equation having a common factor with the similar equation) that one of the functions $\phi(a, \ldots)$ was equal to one of the functions $\phi(b, \ldots)$, which is contrary to the hypothesis. From this it follows that a can be expressed as a rational function of t, and it is the same for the other roots." (See Appendix 1.)

When Poisson read Galois' memoir in 1831, he made a note in the margin next to this† proof saying "The proof of this lemma is insufficient, but it is true by article 100 of the memoir of Lagrange, Berlin, 1771." Galois replied in a note on the manuscript, "*On jugera,*" that is, one will make up one's own mind. Perhaps he meant something like, "That remains to be seen."

It is easy to understand Poisson's position. Galois' proof can be regarded as, at best, a sketch, and therefore is certainly "insufficient" if one is in any doubt as to the correctness of his theory and the accuracy of his reasoning. In his report to the Academy (which formally involved Lacroix, but Lacroix did not even sign it), Poisson said of Galois' memoir as a whole that "We

---

* Galois used the letter V for t.

† In the recent edition [G1] of Galois' writings, the pair of asterisks indicating this note is badly placed, indicating that it pertains to Lemma II rather than Lemma III. As is clear from both the content and from the reproduction of the actual note in the front of the volume, it pertains to Lemma III. See also p. 491 of this edition for confirmation that the proof of Lemma III is the one in question.

have made every effort to understand Mr. Galois' proof. His arguments are not clear enough, nor developed enough, for us to be able to judge their correctness. . . ." He hoped that Galois would improve and amplify his exposition of his work, but concluded that "in the state in which it is now submitted to the Academy, we cannot recommend that you give it your approval."* With the benefit of hindsight, one can see that Galois' ideas were correct and gave great insight into algebraic problems. At the time, confronted with an incomprehensible manuscript and a 19-year-old author who could well be asked to improve on it (and who was in trouble with the police to boot), one might well decide to recommend to one's colleagues that they not endorse it.

Galois' position is also easy to understand. *He* knew that his proofs were correct and that his understanding of algebra far exceeded that of the men who were judging his work. Moreover, he had twice been refused admission to the Ecole Polytechnique, when he surely knew he was better qualified than most who were accepted, and the previous year he had submitted a memoir to the Academy to compete for a prize and it wasn't even considered for the prize because it was lost! After Poisson's rejection of his paper, he became extremely bitter and evidently gave no consideration at all to amplifying his memoir and making it more understandable to the likes of Poisson.

On jugera. Is Galois' proof sufficient? Of course, a proof is in the eye of the beholder, so a flat answer yes or no cannot be given. To some extent the answer depends on what one is allowed to assume as known. However, assuming the existence of a field $K(a, b, c, \ldots)$ in which the given polynomial has n distinct roots, one can amplify Galois' argument slightly to give a convincing proof of his Lemma III and therefore to make his theory quite independent of Lagrange—something that Galois would probably have insisted upon. The argument is as follows.†

Let $F(X, Y)$ be the polynomial obtained from the product of the $(n - 1)!$ expressions $X - \phi(Y, b, c, d, \ldots)$ when the roots $b, c, d, \ldots$ are permuted in all possible ways, when the symmetric polynomials in $b, c, d, \ldots$ are expressed in terms of a (see the end of §30), and when a is replaced by Y. Then $F(t, a) = 0$, as Galois says, and what is to be shown is that this equation can be solved for a in terms of t. Let f be the given polynomial with roots $a, b, c, \ldots$. Let the Euclidean algorithm be used to find the greatest common divisor of $f(Y)$ and $F(t, Y)$ when both are regarded as polynomials in Y with coefficients in the field $K(t)$. This gives $d(Y) = A(Y)F(t, Y) + B(Y)f(Y)$ where d, A, and B are polynomials with coefficients in $K(t)$ and $d(Y)$ is a divisor of both $F(t, Y)$ and $f(Y)$. Since $Y - a$ divides‡ both $F(t, Y)$ and $f(Y)$, it divides

---

*Taton [T1].

† Note added in second printing: For further discussion of this point see Edwards, Harold M., *A Note on Galois Theory,* Archive for History of Exact Sciences, vol. 41 (1990) pp. 163–169. See also Abel's paper "Recherche de la quantité qui satisfait a la fois a deux équations algébriques données" (*Oeuvres* vol. 1, p. 212).

‡ Division of a polynomial $g(Y)$ by $Y - a$ gives $g(Y) = q(Y)(Y - a) + r$ where r is constant. Substitution of a for Y gives $g(a) = r$. Thus $Y - a$ divides $g(Y)$ (that is, $r = 0$) if and only if $g(a) = 0$. This is known as the *Remainder Theorem.*

$d(Y)$. Galois contends that $d(Y)$ has no other factors. Since $f(Y)$ is a constant times $(Y - a)(Y - b)(Y - c)\ldots$, $d(Y)$ is a (nonzero) constant times $Y - a$ times a product of some subset of the factors $Y - b$, $Y - c$, ... If $Y - b$ divides $d(Y)$ then it divides $F(t, Y)$, that is, $F(t, b) = 0$. To prove Galois' contention that $d(Y) = \gamma(Y - a)$ ($\gamma \in K(t)$, $\gamma \neq 0$) it will suffice to prove that $F(t, b) \neq 0$ and, by symmetry, $F(t, c) \neq 0$, ...

The expressions of the elementary symmetric functions of $n - 1$ of the roots in terms of the remaining root—namely $\tau_1 = \sigma_1 - a$, $\tau_2 = \sigma_2 - a\sigma_1 + a^2$, $\tau_3 = \sigma_3 - a\sigma_2 + a^2\sigma_1 - a^3, \ldots$—are independent of the choice of the special root. Therefore $F(X, Y)$ can also be described as the product of all $(n - 1)!$ expressions $X - \phi(Y, a, c, d, \ldots)$ when the roots $a, c, d, \ldots$ are permuted in all possible ways, when the symmetric polynomials in $a, c, d, \ldots$ are expressed in terms of b, and when b is replaced by Y. This shows that $F(t, b)$ is equal to the product of the $(n - 1)!$ terms $t - \phi(b, a, c, \ldots) = \phi(a, b, c, \ldots) - \phi(b, a, c, \ldots)$. Since by assumption these factors are not zero, this shows that $F(t, b) \neq 0$ and consequently that $d(Y) = \gamma(Y - a)$ ($\gamma \in K(t)$, $\gamma \neq 0$).

Setting $Y = 0$ in $d(Y) = A(Y)F(t, Y) + B(Y)f(Y)$ gives $\gamma \cdot (-a) \in K(t)$. Then division by $-\gamma$ gives $a \in K(t)$, as was to be shown. (See Exercise 9 for an application of Galois' method in a specific case.)

Fourth Exercise Set

1. As in §30, let t be a polynomial in n variables $x_1, x_2, \ldots, x_n$, let $t_1, t_2, \ldots, t_k$ be all distinct polynomials in $x_1, x_2, \ldots, x_n$ that can be obtained from t by permutation of the variables, and let $\phi_1, \phi_2, \ldots, \phi_k$ be permutations of the variables such that $t_i = \phi_i t$. Show that if ψ is any permutation of the variables then $\psi = \phi_i g$—that is, first perform the permutation g and then perform the permutation ϕ_i—where $i = 1$ or 2 or $\ldots$ or k and where $gt = t$.

2. Let f be a polynomial with coefficients in a field K. Let $f(x + h)$, which is a polynomial in two variables x and h, be expanded $f(x + h) = A(x) + B(x)h + C(x)h^2 + \ldots$ as a polynomial in h with coefficients that are polynomials in x. Then $A(x) = f(x)$. The coefficient $B(x)$ of h is by definition the *derivative* of $f(x)$, denoted $f'(x)$. Show that $f(x)$ divided by its greatest common divisor with $f'(x)$ has the same roots as $f(x)$ and has only simple roots.

3. Show that if K is a field, if $x_1, x_2, x_3, \ldots$ is an infinite sequence of distinct elements of K, and if $F(A, B, C, \ldots)$ is a nonzero polynomial in n variables $A, B, C, \ldots$ with coefficients in K, then it is possible to select values $A = x_j$, $B = x_k$, $C = x_m, \ldots$ for the variables $A, B, C, \ldots$ from the sequence $x_1, x_2, x_3, \ldots$ so that $F(x_j, x_k, x_m, \ldots) \neq 0$.

4. Prove that a product of nonzero polynomials is nonzero.

5. Describe the field of complex numbers as a simple algebraic extension of the field of real numbers.

6. Find $(3 + 2\sqrt{2})^{-1}$ in the field $\mathbb{Q}(\sqrt{2})$ obtained by adjoining a root $\sqrt{2}$ of $x^2 - 2$ to the field $\mathbb{Q}$.

7. Show that if a, b, c are the roots of an irreducible cubic equation with rational co-efficients then $a - b$ is a Galois resolvent of the equation. [If an equation has a rational root it is not irreducible, nor is it irreducible — by Exercise 2 — if it has multiple roots.]

8. Let $t = a - b$ be the Galois resolvent of an irreducible cubic treated in the preceding exercise. Under the assumption that the irreducible cubic has the special form $x^3 + px + q$ (a change of variable puts any cubic in this form) find explicitly the 6th degree polynomial $F(X)$ of which t is a root. Show that $F(X)$ is reducible if the discriminant $-4p^3 - 27q^2$ is a square. (It will be seen in Exercise 5 of the next set that $F(X)$ is reducible *only if* $-4p^3 - 27q^2$ is a square, in which case it is a product of two irreducible cubics.)

9. Use Galois' method of §37 to express the roots a, b, c of an irreducible cubic $x^3 + px + q = 0$ as rational functions of $t = a - b$. Of the three, c has the simplest expression. Verify that if t is a root of the equation found in Exercise 8 then this rational function c of t is a root of $x^3 + px + q = 0$.

10. In the case of the equation $x^3 - r = 0$, express c as a *polynomial* in t. Since the ratio of any two roots is a cube root of unity, the field $K(t)$ must contain a square root of -3. Find one.

11. Combine Exercises 8 and 9 to express a, b, c, the roots of an irreducible cubic $x^3 + px + q = 0$, as *polynomials* in $t = a - b$.

12. Galois' derivation assumes the existence of a field in which the given equation $f(x) = 0$ has deg f roots. Prove without this assumption that the field $K(t)$, obtained by adjoining a root t of an irreducible factor of the polynomial found in Exercise 8, contains three roots of $x^3 + px + q$ in the sense that $x^3 + px + q = (x - a)(x - b)(x - c)$ where a, b, $c \in K(t)$.

13. The *theorem of the primitive element* states that if r, s, t, ... is a finite set of elements each of which is algebraic over K — that is, each of which is a root of a polynomial with coefficients in K — then $K(r, s, t, ...) = K(w)$ for some w. In other words, there is a rational function w of r, s, t, ... such that each of the elements r, s, t, ... can be expressed as a rational function of w. Show that this theorem is equivalent to the following one: Let f be a polynomial with coefficients in K and, as in the text, let $K(a, b, c, ...)$ be the field of all rational functions in the roots a, b, c, ... of f. Given any subset of the roots a, b, c, ... there is a $u \in K(a, b, c, ...)$ which is a rational function of the roots in the subset and in terms of which each of the roots in the subset can be expressed as a rational function. In short, $K(u) = K(a', b', c', ...)$ where a', b', c', ... run over roots of the given subset of a, b, c, ...

14. Prove the theorem of the primitive element as it is reformulated in the preceding exercise by generalizing Galois' proof of §37. [Van der Waerden has stated [W1] that Galois' proof was the inspiration for his proof of the theorem of the primitive element, although he does not explain the connection.]

15. It is obvious that anything expressible as a polynomial in the n roots a, b, c, ... of an equation $f(x) = 0$ of degree n with coefficients in K can be expressed as a polynomial of degree $< n$ in each of the roots a, b, c, ... Show that in fact it can be expressed as a polynomial of degree $< n$ in a, of degree $< n - 1$ in b, of degree $< n - 2$ in c, ..., and not involving the last root at all.

16. If you are familiar with the notion of an *algebraic integer* in an extension field of $\mathbb{Q}$, you can prove the existence of a Galois resolvent for an equation with rational coefficients and nonzero discriminant D as follows. Let $a_1, a_2, \ldots a_n$ be the roots of the equation. Let m be an integer such that ma_i is an algebraic integer for $i = 1, 2, \ldots, n$. Then $m^{n(n-1)}D$ is an integer. Let $p_1, p_2, \ldots, p_n$ be distinct prime integers which do not divide $m^{n(n-1)}D$. Let A_i be the product of $p_1, p_2, \ldots, p_n$ except p_i and let $t = \sum A_i m a_i$. Then t is an algebraic integer and a Galois resolvent because if $\sum A_i ma_{\sigma(i)} = \sum A_i ma_{\tau(i)}$ then $ma_{\sigma(i)} \equiv ma_{\tau(i)} \bmod p_i$; if $\sigma(i) \neq \tau(i)$ this gives $m^{n(n-1)}D \equiv 0 \bmod p_i$, contrary to assumption. Fill in this sketch.

17. Show that if $f(x) = x^2 + px + q$ has nonzero discriminant then $Aa + Bb$ is a Galois resolvent if and only if $A \neq B$.

18. Let $b(X)$ be a given nonzero polynomial with coefficients in a field K. Suppose it is known that every polynomial $a(X)$ of degree $< k$ can be written in the form $a(X) = q(X)b(X) + r(X)$ where $\deg r < \deg b$. Show that the same is true for polynomials $a(X)$ of degree k, and conclude that it is true for all polynomials $a(X)$. Show also that $q(X)$, $r(X)$ are uniquely determined by $a(X)$, $b(X)$.

19. Describe a computational scheme similar to long division for computing, given polynomials $a(X)$ and $b(X)$ with coefficients in a field K, polynomials $q(X)$ and $r(X)$ with coefficients in the same field and with $a(X) = q(X)b(X) + r(X)$, $\deg r < \deg b$. (Or, write a computer program for doing this in the case $K = \mathbb{Q}$. You may ignore the fact that normal calculations on the computer imperfectly represent $\mathbb{Q}$.)

Basic Galois Theory: The Galois Group

§38   As was seen above, the idea of the Galois resolvent is not a great step beyond Lagrange's work. It is simply the recognition that if one is willing to drop the condition that the equation for t be solvable then there is always a "resolvent" t, that is, a polynomial t in the roots such that every rational function of the roots is a polynomial in t. The great advance which Galois made was to devise a means of analyzing the structure of the field $K(t)$ of rational functions of the roots which enables one, in theory, to determine whether the roots can be expressed in terms of known quantities from K and the operations of addition, subtraction, multiplication, division, and the extraction of roots. The main tool in this analysis was the concept of what is today called the *Galois group* of the field $K(t)$ over K.

Before proceeding to the discussion of the Galois group, it will be useful to review briefly the notation and the notion of the Galois resolvent. Let K be the field of quantities that are assumed to be known, and let $f(x) = 0$ be a polynomial equation of degree n with coefficients in K. Let $a, b, c, \ldots$ be the roots of $f(x) = 0$. (Galois and his predecessors, going all the way back to Girard 200 years earlier, seem to have taken it for granted that it was meaningful to talk about the roots of $f(x)$ and that there were precisely n roots. They do not seem to have worried about where or what these roots were. This question will be addressed in §49 *et seq.*) It will be assumed that the roots $a, b, c, \ldots$ are *distinct*, a restriction which does not cost any generality because there is a simple rechnique for replacing an equation $f(x) = 0$ with multiple roots with another equation with the same roots, all occurring once. (See §31.) Let $K(a, b, c, \ldots)$ denote the field of all elements that can be expressed as rational functions of the roots $a, b, c, \ldots$. Then there is an element $t \in K(a, b, c, \ldots)$, called a *Galois resolvent*, with the property that any given element of $K(a, b, c, \ldots)$ can be expressed as a polynomial in t with coefficients in K. In fact, t can be taken to have the form $t = Aa + Bb + Cc + \cdots$ where $A, B, C, \ldots$ are integers. Let $F(X)$ be the polynomial $F(X) = \prod (X - ASa - BSb - CSc - \cdots)$ with coefficients in K, where the product is over all $n!$ permutations S of the roots $a, b, c, \ldots$, and let

$$F(X) = G_1(X)G_2(X) \cdots G_k(X)$$

be a decomposition of F into irreducible factors. (See §61 below.) Then $F(t) = 0$, and, consequently, t is a root of one of the polynomials G_i, say $G_1(t) = 0$. Then the field $K(t)$, which is isomorphic to $K(a, b, c, \ldots)$, can be represented very concretely as the simple algebraic extension of K obtained by adjoining a root of the irreducible polynomial G_1 (see §§34–36).

§39 Most readers probably have some acquaintance with the notion of a finite group. Galois was the first[*] to introduce it, and he did so in the following very concrete way.

An *arrangement* of the n roots $a, b, c, \ldots$ is simply a listing of them in some order. A *substitution* of them is a one-to-one onto mapping of the roots to themselves. Two arrangements can be regarded as representing a substitution, namely, the substitution which transforms the first arrangement into the second. For example, the ordered pair $(abcdef, bacefd)$ represents the substitution which interchanges a and b, leaves c fixed, and performs the cyclic substitution $d \to e \to f \to d$ of d, e, and f. A *presentation of a group*[†] is a list of arrangements of $a, b, c, \ldots$ with the property that the set of substitutions which transforms the first arrangement in the list to each of the others is the same as the set of substitutions which carry any other arrangement in the list to the remaining ones. For example,

$$
\begin{array}{cccccc}
a & b & c & d & e & f \\
b & a & c & e & f & d \\
a & b & c & f & d & e \\
b & a & c & d & e & f \\
a & b & c & e & f & d \\
b & a & c & f & d & e
\end{array}
$$

[*] Note add in second printing: Although Galois was the first to use the word "group" in this way, and although he was the first to represent a finite group in the way that is described in this section, Cauchy certainly dealt with finite groups (see, for example, "Sur le nombre des valeurs qu'une fonction peut acquerir, lorsqu'on y permute de toutes les manieres possible les quantités qu'elle renferme") and to some extent so did Lagrange and Ruffini.

[†] The term "presentation of a group" is used differently in the literature of group theory.

is a presentation of a group. If the columns are rearranged so that the 1st row is in the order of the 4th row, say, the table becomes

$$
\begin{array}{cccccc}
b & a & c & d & e & f \\
a & b & c & e & f & d \\
b & a & c & f & d & e \\
a & b & c & d & e & f \\
b & a & c & e & f & d \\
a & b & c & f & d & e.
\end{array}
$$

On the one hand, rearranging the columns does not alter the substitutions; therefore, the two tables describe the same substitutions. On the other hand, the rows of the second table are a rearrangement of those of the first, so these substitutions are the substitutions which carry the 4th row of the first table to the remaining rows. In order to show that the first table (and therefore the second table) is a presentation of a group, one would have to show that what was just shown to be true of the 4th row is true of the other rows also.

A *group* of substitutions is a set of substitutions that can be obtained in this way from a presentation of a group, including the identity substitution which carries the first arrangement to itself. Galois observes that *a set of substitutions is a group if and only if it is closed under composition*, that is, if and only if for any two substitutions S and T in the set, their composition ST is in the set. The proof of this fact is left as an exercise to the reader (Exercise 1). In the example above, let S be the substitution which carries the 1st row to the 4th (S is the interchange $a \leftrightarrow b$) and let T be the one which carries the 1st row to the 5th (T is the cyclic substitution $d \rightarrow e \rightarrow f \rightarrow d$). Then S and T commute, $S^2 = T^3 = $ identity, and the six rows of the table correspond to the identity substitution, ST, T^2, S, T, and ST^2. Since this set of substitutions is closed under composition, it follows that the table is a presentation of a group.

The terminology introduced here is an adaptation of Galois' terminology intended to make Galois' memoir easily accessible to a modern reader. Galois calls an arrangement a "permutation." Since the word "permutation" usually means—and even in Galois' day usually meant—what is called a substitution above and in Galois' memoir, it seems better to avoid this word altogether. Galois is not very clear as to just what he means by a "group." For the most part he seems to mean what is called a presentation of a group above,* but he also uses the phrase "group of substitutions", by which he surely meant what is called a group above. Since the modern meaning of the word group is so firmly established, it would be foolish to use it in any other way here. More exactly, what is called a group above is, of course, a subgroup of the group of all substitutions (permutations) of the n roots $a, b, c, \dots$. This is the only kind of group that will be considered here.

---

* See, for example, his Proposition I.

§40 A *subgroup* of a group is simply a group which is contained in another. In terms of a presentation of the group, a subgroup partitions the presentation into a number of presentations of the subgroup. For example, if the group is the group of all substitutions of the three letters a, b, c and the subgroup is the identity and the interchange $a \leftrightarrow b$ then the three presentations

$$
\begin{array}{ccc}
a \quad b \quad c & a \quad c \quad b & c \quad a \quad b \\
b \quad a \quad c & b \quad c \quad a & c \quad b \quad a
\end{array}
\tag{1}
$$

of the subgroup taken together give a presentation of the whole group. The number of presentations of the subgroup in a presentation of the group is called the *index* of the subgroup. To say the same thing another way, the number of substitutions in the subgroup must divide the number of substitutions in the group, and the quotient is called the *index* of the subgroup in the group. These facts too are left to the reader to prove (Exercise 2).

Galois recognized the special importance of what are today called *normal* subgroups, although he described them in a manner that is unfamiliar today. As was just noted, a subgroup of index k of a group gives a partition of a presentation of the group into k presentations of the subgroup. Galois singled out those subgroups with the property that the various presentations of the subgroup differ from one another by the application of a single substitution.* For example, the subgroup presented in (1) above does not have this property because no substitution of the letters in the first presentation gives the second presentation. (No matter how the names of the letters in the first presentation are changed, the letter in the 3rd column is the same in both rows, and this is not true of the second presentation.) A subgroup which does have this property is

$$
\begin{array}{ccc}
a \quad b \quad c & \quad & c \quad b \quad a \\
b \quad c \quad a & \quad & b \quad a \quad c \\
c \quad a \quad b & \quad & a \quad c \quad b.
\end{array}
\tag{2}
$$

(The group is all six substitutions of three letters. The subgroup of three substitutions has these two presentations. The interchange of a and c carries the one presentation of the subgroup to the other, as does any interchange.) Another such subgroup is Galois' example

$$
\begin{array}{cccc}
a \quad b \quad c \quad d & a \quad c \quad d \quad b & a \quad d \quad b \quad c \\
b \quad a \quad d \quad c & c \quad a \quad b \quad d & d \quad a \quad c \quad b \\
c \quad d \quad a \quad b & d \quad b \quad a \quad c & b \quad c \quad a \quad d \\
d \quad c \quad b \quad a & b \quad d \quad c \quad a & c \quad b \quad d \quad a.
\end{array}
\tag{3}
$$

* For purposes of this definition, it is necessary to regard two presentations as being the *same* whenever they consist of the same rows, even if they list these rows in a different order.

(The group contains twelve substitutions of four letters. The subgroup contains four substitutions. It should be checked that the three subpresentations present the same group. Clearly, the substitution which carries the 1st row of any one of these three subpresentations to the 1st row of any other also carries the following rows of the one to the following rows of the other.) It is a simple exercise (Exercise 3) to prove that a subgroup has this property if and only if it is true that, for any T in the group, and any S in the subgroup, the substitution $T^{-1}ST$ is in the subgroup. Such a subgroup is called a *normal* subgroup.

§41   The *Galois group* of the given equation $f(x) = 0$ with roots $a, b, c, \ldots$ is the group with the following presentation. Let each of the roots $a, b, c, \ldots$ be expressed in terms of the Galois resolvent t, say $a = \phi_a(t)$, $b = \phi_b(t)$, $c = \phi_c(t), \ldots$, where the ϕ's are polynomials with coefficients in K. As above, let $F(X)$ be the polynomial of degree $n!$ whose roots are the $n!$ distinct elements $ASa + BSb + CSc + \cdots$ of $K(a, b, c, \ldots)$, where $t = Aa + Bb + Cc + \cdots$ and where S is one of the $n!$ substitutions of $a, b, c, \ldots$. Finally, let $G(X)$ be an irreducible (over K) factor of $F(X)$ of which t is a root. (See §38.) The *conjugates* $t', t'', \ldots$ of t (over K) are the other roots of $G(X)$. The *Galois group* of $f(x) = 0$ is presented by

$$\phi_a(t) \quad \phi_b(t) \quad \phi_c(t) \ldots,$$
$$\phi_a(t') \quad \phi_b(t') \quad \phi_c(t') \ldots, \tag{1}$$
$$\phi_a(t'') \quad \phi_b(t'') \quad \phi_c(t'') \ldots,$$
$$\cdots.$$

The number of rows in the table—and therefore the number of elements in the group—is equal to the number of roots $t, t', t'', \ldots$ of G, which is equal to deg G. (Note that the conjugates $t', t'', \ldots$ are all roots of F and are therefore of the form $ASa + BSb + CSc + \cdots$ where S is a substitution of $a, b, c, \ldots$.)

In order to justify this definition of the Galois group it must be shown that:

(A) for any conjugate t' of t, the elements $\phi_a(t'), \phi_b(t'), \phi_c(t'), \ldots$ are an arrangement of the 1st row $\phi_a(t) = a, \phi_b(t) = b, \phi_c(t) = c, \ldots$;

(B) the arrangements in the table present a group; and

(C) the group of substitutions of the roots $a, b, c, \ldots$ presented by the table is independent of the choice of the Galois resolvent t.

In his proof of (A), and in several later proofs, Galois makes use of a simple and very basic lemma:

Lemma I. *If $g(X)$ and $h(X)$ are polynomials with coefficients in a given field K, if $g(X)$ is irreducible, and if $g(X)$ and $h(X)$ have a common root, then $g(X)$ divides $h(X)$.*

PROOF. Let the Euclidean algorithm be used to write a common divisor $d(X)$ of $g(X)$ and $h(X)$ in the form $d(X) = A(X)g(X) + B(X)h(X)$ where $A(X)$ and $B(X)$ are polynomials with coefficients in K. If the common root r of $g(X)$ and $h(X)$ is substituted into this equation the result is $d(r) = 0$. Therefore $d(X)$ has degree > 0. Since $d(X)$ divides $g(X)$ and $g(X)$ is irreducible, it follows that $d(X)$ is a nonzero element of K times $g(X)$. Since $d(X)$ divides $h(X)$, it follows that $g(X)$ divides $h(X)$, as was to be shown. $\square$

Using this lemma, Galois proves (A) as follows. Since $f(\phi_a(X))$ is a polynomial in X with coefficients in K, and since it has the root t in common with the irreducible polynomial $G(X)$, by Lemma I it is divisible by $G(X)$ and every root of $G(X)$ is a root of $f(\phi_a(X))$. Similarly, every root of $G(X)$ is a root of $f(\phi_b(X)), f(\phi_c(X)), \ldots$ This shows that the entries in the presentation (1) of the Galois group are all roots of the equation $f(x) = 0$. If two entries in the same row are equal, there is an equation of the form $\phi_d(t') = \phi_e(t')$, where d and e are roots of $f(x) = 0$ and t' is a root of $G(X)$. By Lemma I, $G(X)$ divides $\phi_d(X) - \phi_e(X)$ and therefore t is a root of $\phi_d(X) - \phi_e(X)$. Thus $d = \phi_d(t) = \phi_e(t) = e$. Thus, each row of the presentation of the Galois group includes each root of $f(x) = 0$ at most once, and (A) follows.

Galois does not address (B) and (C) directly, but it is clear that he saw them as consequences of his Proposition I. His statement of this Proposition requires rather a lot of explanation about the "substitutions" it is talking about and the type of "invariance" it means. Therefore a reformulation of it seems preferable to a quotation. I believe that this reformulation is faithful to his meaning.*

Proposition I. *Let* $\Psi(U, V, W, \ldots)$ *be a polynomial in n variables with coefficients in K. Let* $\Psi_{t'}$ *be the element of* $K(a, b, c, \ldots)$ *obtained by setting* $U = \phi_a(t')$, $V = \phi_b(t')$, $W = \phi_c(t'), \ldots$ *in* Ψ. *In other words, for U, V, W, \ldots one substitutes the arrangement of the roots given in the row of the above table* (1) *corresponding to t'. Similarly, let* $\Psi_t, \Psi_{t''}, \Psi_{t'''}, \ldots$ *be defined by substituting the corresponding row. Then* Ψ_t *is in K if and only if the elements* $\Psi_t, \Psi_{t'}, \Psi_{t''}, \ldots$ *are all equal. Loosely speaking,* $\Psi(a, b, c, \ldots)$ *is a known quantity if and only if it is invariant under all the substitutions of the Galois group.*

PROOF. Substitute $U = \phi_a(X)$, $V = \phi_b(X)$, $W = \phi_c(X), \ldots$ in $\Psi(U, V, W, \ldots)$ to obtain a polynomial $\Psi^*(X)$ with coefficients in K. If Ψ_t is in K then

$$\Psi^*(X) - \Psi_t$$

is a polynomial with coefficients in K of which t is a root. By Lemma I, $G(X)$ divides $\Psi^*(X) - \Psi_t$, and therefore t' is a root of this polynomial when t' is any root of G. Thus $\Psi_{t'} - \Psi_t = 0$ and, since t' was arbitrary, $\Psi_{t'}, \Psi_{t''}, \ldots$

---

* For Galois' statement, see Appendix 1.

are all equal to Ψ_t. Conversely, if $\Psi_t, \Psi_{t'}, \Psi_{t''}, \ldots$ are all equal and if there are k of them then

$$\Psi_t = \frac{1}{k}[\Psi_t + \Psi_{t'} + \Psi_{t''} + \cdots] = \frac{1}{k}[\Psi^*(t) + \Psi^*(t') + \Psi^*(t'') + \cdots].$$

This is a symmetric polynomial in the roots $t, t', t'', \ldots$ of $G(X)$ and can therefore be expressed in terms of the coefficients of $G(X)$. Since these coefficients are in K, it follows that Ψ_t is in K, as was to be shown. $\square$

In order to deduce (B) and (C) from Proposition I, it is helpful to observe:

Corollary. *Let t be a Galois resolvent, let t' be one of its conjugates, and let S be the substitution of the roots which carries the row of the above table* (1) *corresponding to t to the row corresponding to t'. In short, let $S(a) = \phi_a(t')$, $S(b) = \phi_b(t'), S(c) = \phi_c(t'), \ldots$. Then S can be extended to an automorphism of the entire field $K(a, b, c, \ldots)$ over K, that is, a function S from $K(a, b, c, \ldots)$ to itself which is the identity map on K, which carries sums to sums, products to products, and agrees with the original S for the roots $a, b, c, \ldots$.*

PROOF. An element of $K(a, b, c, \ldots)$ can be written as a polynomial in t. Since t can be assumed to be a linear polynomial in $a, b, c, \ldots$, an element of $K(a, b, c, \ldots)$ can therefore be written as a polynomial in $a, b, c, \ldots$. In other words, any given element of $K(a, b, c, \ldots)$ can be written in the form $\Psi(a, b, c, \ldots)$ where Ψ is a polynomial in n variables with coefficients in K. If there is an automorphism extending S, it clearly must carry this element to $\Psi(Sa, Sb, Sc, \ldots)$. One could *define* the effect of the automorphism on the given element to be $\Psi(Sa, Sb, Sc, \ldots)$ if it were known that this definition was independent of the choice of the representation $\Psi(a, b, c, \ldots)$ of the given element. But this follows immediately from Proposition I because if $\Psi(a, b, c, \ldots) = \Phi(a, b, c, \ldots)$ then their difference is 0, which lies in K, and by Proposition I the substitution S of the roots leaves their difference unchanged, that is, $\Psi(Sa, Sb, Sc, \ldots) = \Phi(Sa, Sb, Sc, \ldots)$, as was to be shown. $\square$

Corollary. *Let t be a Galois resolvent, let t' be one of its conjugates, and let S be the corresponding substitution of the roots as in the previous corollary. Let s be any other Galois resolvent of the same equation. Then there is a conjugate s' of s such that S is equal to the substitution of the roots corresponding to s and s'.*

PROOF. Since s is in $K(a, b, c, \ldots)$, the extension of S defined in the preceding corollary applies to s. Define s' to be $S(s)$. If $H(x)$ is an irreducible polynomial of which s is a root, then application of S to $H(s) = 0$ gives $H(s') = 0$, so that s' is a conjugate of s. Now if $a = \psi_a(s)$, where ψ_a is a polynomial with coefficients in K, then application of S to both sides gives $S(a) = \psi_a(Ss) = \psi_a(s')$. Similarly, $S(b) = \psi_b(s'), S(c) = \psi_c(s'), \ldots$, as was to be shown. $\square$

Now (B) follows immediately from the last corollary because this corollary shows that if t' is any conjugate of t then (set $s = t'$) the substitutions of the roots that result from changing t to one of its conjugates are the same as those that result from changing t' to one of its conjugates. (The conjugates of t coincide with those of t' because in both cases the conjugates are the roots of G.) Since (C) follows even more immediately from the last corollary, this completes the justification of the above definition of the Galois group of the equation $f(x) = 0$ as a group of substitutions of the roots of the equation. Moreover, it shows that the elements of the Galois group can also be regarded as automorphisms of the field $K(a, b, c, \ldots)$ over K—a fact which is not stated explicitly by Galois, but which is fundamental in most modern formulations of his theory.

Examples

§42 Galois gives two examples of Galois groups of equations. The first is what he calls an "algebraic" equation, by which he means an equation in which the coefficients are indeterminates rather than numbers. For such equations, he says that the group is the set of all $n!$ substitutions of the roots "because in this case the symmetric functions [of the roots] are the only ones that can be determined rationally," that is, are the only ones in K. In other words, *only* symmetric functions of the roots can be expressed in terms of the coefficients and therefore, by Proposition I, the group must contain all substitutions. This is rather plausible, but the proof is far from obvious. (See §67.)

His other example is the group of the equation $x^{p-1} + x^{p-2} + \cdots + x + 1 = 0$ satisfied by the primitive pth roots of unity where p is prime. Here if a is any root of the equation the other roots are $a^2, a^3, \ldots, a^{p-1}$. If S is any element of the Galois group, then $S(a) = a^k$ for some $k = 1, 2, \ldots, p - 1$ (because $S(a)$ is a root). Thus the effect of S on any root is known because $S(a^j) = S(a)^j = a^{jk}$ (S is an automorphism). The substitutions $S(a^j) = a^{jk}$ are more conveniently described by choosing a primitive root g mod p (see §24) and setting $a_0 = a, a_1 = a_0^g, a_2 = a_1^g, a_3 = a_2^g, \ldots$. Then S carries a_i to $a_{i+\mu}$, where μ is defined by $S(a_0) = a_\mu$, because $S(a_1) = S(a_0^g) = S(a_0)^g = a_\mu^g = a_{\mu+1}$, $S(a_2) = a_{\mu+2}, \ldots$. Thus the group of this equation is contained in the group presented by

$$
\begin{array}{ccccc}
a_0 & a_1 & a_2 & \cdots & a_{p-2} \\
a_1 & a_2 & a_3 & \cdots & a_0 \\
a_2 & a_3 & a_4 & \cdots & a_1 \\
\vdots & \vdots & \vdots & & \vdots \\
a_{p-2} & a_0 & a_1 & \cdots & a_{p-3}.
\end{array}
\tag{1}
$$

The group depends on the field K of known quantities. Galois, clearly considering the case where K is the field of rational numbers, states, without proof, that the group is the set of *all* $p - 1$ substitutions presented in the table above. The proof of this fact uses a theorem of Gauss (Exercise 10).

Another example of interest is an equation of the form $x^p = k$ where the known quantities K include k and a primitive pth root of unity α (with p prime). If a is any root of the equation $x^p - k = 0$ then the other roots are αa, $\alpha^2 a, \ldots, \alpha^{p-1}a$. If S is any substitution of the Galois group then $S(a) = \alpha^j a$ for some j and all values of S are given by $S(\alpha^k a) = \alpha^k S(a) = \alpha^{k+j}a$ (because S is an automorphism that leaves the elements of K fixed). Thus

$$
\begin{array}{ccccc}
a & \alpha a & \alpha^2 a & \ldots & \alpha^{p-1}a \\[4pt]
\alpha a & \alpha^2 a & \alpha^3 a & \cdots & a \\[4pt]
\alpha^2 a & \alpha^3 a & \alpha^4 a & \cdots & \alpha a \\[4pt]
& & & \cdots & \\
\vdots & \vdots & \vdots & \cdots & \vdots \\
& & & \cdots & \\[4pt]
\alpha^{p-1}a & a & \alpha a & \ldots & \alpha^{p-2}a
\end{array}
\tag{2}
$$

is a presentation of a group containing the Galois group. Since this is a group with p elements, and since the number of elements in a subgroup must therefore divide p, the Galois group is either the entire group presented above or it is the group containing the identity alone. In the latter case, by Proposition I, every element of $K(a, b, c, \ldots)$ is in K and the equation already has p roots in K. Otherwise the Galois group is the group presented above. A simple consequence of this observation is that if k does not have a pth root in K and if K contains a primitive pth root of unity then the polynomial $x^p - k$ is irreducible over K (see Exercise 4).

As a final example, consider the equation $x^4 + 1 = 0$ (the equation for the primitive 8th roots of unity). If a is any root then so are a^3, a^5, and a^7, and the group of the equation is easily seen to be contained in the group presented by

$$
\begin{array}{cccc}
a & a^3 & a^5 & a^7 \\[4pt]
a^3 & a & a^7 & a^5 \\[4pt]
a^5 & a^7 & a & a^3 \\[4pt]
a^7 & a^5 & a^3 & a \,.
\end{array}
$$

As before, if K is the field of rational numbers, then the Galois group of the equation is the entire group of four elements presented above (Exercise 11).

For explicit presentations of the Galois groups of cubic equations, see Exercises 6 and 7.

Fifth Exercise Set

1. Show that a set of substitutions of n objects $a, b, c, \ldots$ is a group if and only if it is closed under composition.

2. Show that a subgroup H of a group G partitions a presentation of G into presentations of H and, in particular, that the number of elements in H divides the number of elements in G. [Assume a presentation of G in which no two arrangements are equal, so that the number of arrangements is equal to the number of elements in G.]

3. Show that a subgroup H of a group G is normal in the sense defined in §40 if and only if, for every S in H and T in G, $T^{-1}ST$ is in H.

4. Show that the Galois group acts *transitively* on the roots of a factor of f if and only if the factor is irreducible over K. (To say that the group acts transitively on the roots means that given any two roots a and b there is a substitution of the group which carries a to b.)

5. Let $f(X)$ be an irreducible cubic equation. Show that the Galois group of f is either the full group of all six substitutions of the roots a, b, c of f or it is the three element normal subgroup presented in (2) of §40. Show that the latter case occurs if and only if the discriminant of f is a square.

6. Let $x^3 + px + q$ be an irreducible cubic whose discriminant is a square, say $P^2 = -4p^3 - 27q^2$. Let $K(t)$ be obtained by adjoining a root t of $t^3 + 3pt + P = 0$. Give explicit second degree polynomials a, b, c in t with coefficients in K which are roots of $x^3 + px + q = 0$ in $K(t)$. Show that $t' = b - c$ and $t'' = c - a$ are conjugates of t and find the Galois group.

7. In the case where the discriminant of the irreducible cubic $x^3 + px + q$ is not a square, give the explicit factorization of $F(X)$ (Exercise 8 of the Fourth Set) into linear factors over $K(t)$.

8. Let $f(x)$ be an irreducible cubic whose Galois group has three elements. Then every element of $K(t)$ can be written as a polynomial of degree < 3 in t with coefficients in K. In particular, $1, a$, and a^2 can be so expressed, where a is any root of f. Show, without computation, that these equations can be solved for t as a polynomial in a. Thus the other roots of f can be expressed as polynomials in a. Find these expressions, assuming $\sigma_1 = 0$. Apply them in the cases $f(x) = x^3 - 3x + 1$ and $f(x) = x^3 - 2$ to express the remaining two roots in terms of a given one.

9. Show that the polynomial $F(X)$ of §38 has the property that all of its irreducible factors have the same degree.

10. Gauss proved that, for prime integers p, the polynomial $X^{p-1} + X^{p-2} + \cdots + X + 1$ is irreducible over the rational field $\mathbb{Q}$. A proof of this fact will be given in §64. Making use of it, prove that the Galois group of $(x^p - 1)/(x - 1) = 0$ is the group Galois described (§42).

11. Prove that the Galois group of $x^4 + 1 = 0$ over $\mathbb{Q}$ is the group presented in §42.

12. Find the Galois group of $x^8 - 1 = 0$ over $\mathbb{Q}$.

13. Show that the subgroup presented in Galois' example in (3) of §40 is normal.

14. A group is called *cyclic* if it contains a substitution S with the property that every element of the group is a power of S. For example, the groups presented in (1) and (2) of §42 are cyclic. Show that every subgroup of a cyclic group is cyclic.

15. Show that a subgroup of index 2 is always normal.

16. Let K be a field which contains a pth root of unity $\alpha \neq 1$, where p is prime. Show that, as stated in §42, if a and b are any two roots of $x^p - k = 0$ ($k \in K$) then $b = \alpha^j a$ for some integer j.

17. Show that the automorphism of $K(a, b, c, \ldots)$ over K described in the first corollary of §41 carries t to t'.

18. Let $a, b, c, \ldots$ be the roots of an equation $f(x) = 0$ and let t be any primitive element of the field $K(a, b, c, \ldots)$ (see Exercise 13 of the Fourth Set). Show that the irreducible equation with coefficients in K of which t is a root has degree equal to the number of elements in the Galois group of $f(x) = 0$ over K, that it splits into linear factors over $K(t)$, and that Galois' presentation of the Galois group, (1) of §41, is valid for t (even though t is not necessarily a Galois resolvent).

Basic Galois Theory: The Groups of Solvable Equations

§43   The essence of Galois' achievement was to *determine the conditions imposed on the Galois group of an equation by the assumption that the equation is solvable by radicals*. The first step in doing this, naturally, is to state very explicitly what it means to say that an equation is solvable by radicals.

It will be simplest to assume a strong meaning for "solvable by radicals," namely, that *all* roots (not just one root) of the equation can be expressed in terms of known quantities and the operations of addition, subtraction, multiplication, division by nonzero quantities, and the extraction of roots. In fact, it can be shown (Exercise 4) that if one root of an irreducible equation can be expressed in this way then all can, but this is a fine point that can wait.

If the expression of a root involves taking a jth root and j is not prime, then $j = j_1 j_2$ where j_1 and j_2 are smaller than j, and instead of taking a jth root one can take a j_1th root of a j_2th root. If j_1 or j_2 is not prime, it can be further decomposed, until in the end the expressions of the roots of $f(x) = 0$ involve taking roots of *prime order* only. Moreover, since, as Gauss showed,* the pth roots of unity for any prime p can be expressed in terms of radicals,

---

* As was noted in §24, Gauss's proof was in fact incomplete because he did not prove Lemma 2. It will be convenient to overlook this incompleteness for now and to take it for granted that pth roots of unity can be expressed in terms of radicals. As will be shown in §65, Galois theory can be used to prove this theorem by an argument different from Gauss's. (See also Exercise 5.)

use can be made of pth roots of unity in a solution by radicals. This means that, whenever a pth root of a known quantity is taken, all pth roots of the quantity are known, because all others are obtained from any one by multiplying by pth roots of unity.

The four arithmetic operations can be performed on known quantities without leaving K—that is the definition of a field. However, if the solution calls for taking a pth root, say of a quantity k of K, then although it is possible that there is a pth root of k in K, most likely it will not be possible to do this within K and the realm of known quantities will have to be extended to include the needed pth root of k. Galois called such an extension of the known quantities the *adjunction* of a pth root, and this name has been used ever since. The result of an adjunction is a new, larger field $K' \supset K$ whose elements can all be expressed rationally in terms of $\sqrt[p]{k}$ and quantities in K. (The theorem on simple algebraic extensions can be applied to obtain such a field K' provided $x^p - k$ is shown to be irreducible. It was noted in §42 that this is the case when K contains primitive pth roots of unity. It is even true—see Exercise 6 of the Eighth Set—without this assumption. However, Galois ignored the technicality of constructing K', and consideration of it here will be postponed to §62.) If the solution involves the extraction of another root not already contained in K', then it will be necessary to perform another adjunction $K'' \supset K'$, say of a p_1th root of an element k_1 of K', and so on, until one arrives at a field $K^{(\mu)}$ which contains all the quantities that are involved in the solution of the equation.

Thus, if the given equation $f(x) = 0$ is solvable by radicals, there is a sequence of fields $K \subset K' \subset K'' \subset \cdots \subset K^{(\mu)}$, which starts with the known quantities K, adjoins at the ith stage a p_{i-1}th root of an element k_{i-1} of $K^{(i-1)}$ to $K^{(i-1)}$ to obtain $K^{(i)}$, and ends with a field $K^{(\mu)}$ which contains n roots $a, b, c, \ldots$ of the given equation $f(x) = 0$. Moreover, it can be assumed that $K^{(i-1)}$ contains primitive p_{i-1}th roots of unity.

It is clear that if the field K of known quantities is extended then the Galois group either remains the same or is reduced to a subgroup. This follows from the correspondence between the rows of the presentation of the Galois group and the roots $t, t', t'', \ldots$ of the irreducible factor $G(X)$ of $F(X)$ of which t is a root; if K is enlarged then $G(X)$ may no longer be irreducible, and the roots of the irreducible factor of $F(X)$ of which t is a root may be a proper subset of $t, t', t'', \ldots$, in which case the new Galois group will be a proper subgroup of the old one. Thus, as the field increases $K \subset K' \subset K'' \subset \cdots \subset K^{(\mu)}$, the Galois group decreases. In the end, the roots $a, b, c, \ldots$ have become "known" quantities—that is, quantities in $K^{(\mu)}$—and, by Proposition I, must all be left unchanged by all substitutions of the Galois group. Thus, by the time $K^{(\mu)}$ is reached, the Galois group has been reduced to the identity substitution.

The main step in Galois' analysis of solvable equations is to study the way the group of an equation can be reduced by the adjunction of a pth root of a known quantity when the pth roots of unity are known.

§44 Galois states in his Propositions II and III the main facts about the way in which the Galois group is reduced when the field of known quantities K is extended. The only case that will be needed in what follows is the one in which K is extended by the adjunction of a pth root when it already has pth roots of unity. For Galois' more general propositions see Exercises 6 and 7.

Proposition. *Consider, as above, the Galois group of an equation $f(x) = 0$ over the field K. Let p be a prime, let K contain pth roots of unity, and let $K' \supset K$ be the extension of K obtained by adjoining the pth root of an element k of K. Then the new Galois group (that of $f(x) = 0$ over K') is either the same as the old one or it is a normal subgroup of index p.*

PROOF. As above, let t be a Galois resolvent of $f(x) = 0$, let $F(X) = \prod (X - St)$ be the polynomial of degree $n!$ with coefficients in K of which the $n!$ distinct versions of t in $K(a, b, c, \ldots)$ are roots, and let $G(X)$ be the irreducible factor of $F(X)$ over K of which t is a root. Moreover, let $F(X)$ be factored into irreducible factors over K' and let $H(X)$ be the factor of which t is a root. Since $K' \supset K$, Galois' Lemma I implies that $H(X)$ divides $G(X)$. A presentation of the old Galois group (that of $f(x) = 0$ over K) is obtained by expressing the roots of $f(x) = 0$ as rational functions of t with coefficients in K—say $a = \phi_a(t)$, $b = \phi_b(t)$, $c = \phi_c(t)$, $\ldots$—and listing the arrangements $\phi_a(t')$, $\phi_b(t')$, $\phi_c(t')$, $\ldots$ of $a, b, c, \ldots$ as t' ranges over the deg G distinct roots of $G(X) = 0$. Since $K' \supset K$, a presentation of the new group (that of $f(x) = 0$ over K') is obtained, using the same ϕ's, when t' is subjected to the stronger condition that it be a root of $H(X)$. Thus, the new group is the subgroup of the old group presented by the deg H rows of the presentation of the old group corresponding to roots of $H(X)$ (a subset of the roots of $G(X)$). It is to be shown that either deg $G = $ deg H, or deg $G = p$ deg H and the subgroup is a normal one.

Since $H(X)$ has coefficients in $K' = K(r)$, where r is a pth root of k, and since every element of K' can be expressed as a polynomial in r with coefficients in K, $H(X)$ can be written in the form $H(X) = H(X, r)$ where $H(X, Y)$ is a polynomial in two variables with coefficients in K. Given such a polynomial $H(X, Y)$, consider the polynomial

$$h(X) = H(X, r)H(X, \alpha r)H(X, \alpha^2 r) \cdots H(X, \alpha^{p-1}r),$$

where $\alpha \in K$ is one of the primitive pth roots of unity assumed to exist in K. Then h is a polynomial in X with coefficients that are symmetric polynomials in the p roots $r, \alpha r, \alpha^2 r, \ldots, \alpha^{p-1}r$ of the auxiliary equation $Y^p - k = 0$. By the fundamental theorem on symmetric functions, it follows that the coefficients of h are in K.

On the other hand, since $G(X)$ has coefficients that are in K', Galois' Lemma I implies that $H(X, r)$ divides $G(X)$ with a quotient with coefficients in $K(r)$, say $G(X) = H(X, r)Q(X, r)$.

Lemma. *If $U(X, r) = V(X, r)W(X, r)$ where U, V, and W are polynomials in two variables with coefficients in K (so that $U(X, r)$, $V(X, r)$, and $W(X, r)$ are polynomials in one variable X with coefficients in $K(r) = K'$) then $U(X, \alpha^i r) = V(X, \alpha^i r)W(X, \alpha^i r)$ for all $i = 1, 2, \ldots$.*

PROOF OF LEMMA. The polynomial $U(X, Y) - V(X, Y)W(X, Y)$ can be expanded in the form $\Phi_v(Y)X^v + \Phi_{v-1}(Y)X^{v-1} + \cdots + \Phi_0(Y)$. Substitution of r for Y gives $0 = \Phi_v(r)X^v + \Phi_{v-1}(r)X^{v-1} + \cdots + \Phi_0(r)$. Therefore r is a root of each of the polynomials $\Phi_v(Y)$, $\Phi_{v-1}(Y), \ldots, \Phi_0(Y)$ with coefficients in K. Since $Y^p - k$ is irreducible over K (see §42), Galois' Lemma I implies that it divides all the Φ's. Therefore $U(X, Y) = V(X, Y)W(X, Y) + (Y^r - k)Q(X, Y)$, where $Q(X, Y)$ is a polynomial in two variables with coefficients in K. Substitution of $\alpha^i r$ for Y then gives the desired conclusion.   □

Therefore $G(X) = H(X, \alpha^i r)Q(X, \alpha^i r)$ for $i = 1, 2, \ldots, p$. Multiplication of these p equations gives $G(X)^p = h(X)q(X)$ where $q(X)$ has coefficients in K. Galois' Lemma I can be used to divide this equation p times by $G(x)$ to find $1 = [h(X)/G(X)^j] \cdot [q(X)/G(X)^{p-j}]$. Therefore $h(X) = \text{const.}\ G(X)^j$ for some integer j, where the constant is a nonzero element of K. Comparison of the degrees of the polynomials on the two sides of this equation gives $p \cdot \deg H = j \cdot \deg G$. Since $\deg G/\deg H$ is the index of the new Galois group as a subgroup of the old, this shows that the index divides p (with quotient j). Since p is prime, it follows that the index is 1 or p, as was to be shown. All that remains to be shown is that when the index is p the new group is a *normal* subgroup of the old.

When $\deg G/\deg H = p$, j in the above equation is 1, that is,

$$\text{const.}\ G(X) = H(X, r)H(X, \alpha r) \cdots H(X, \alpha^{p-1}r). \tag{1}$$

Just as the deg H arrangements $\phi_a(t')$, $\phi_b(t')$, $\phi_c(t'), \ldots$ of a, b, $c, \ldots$ corresponding to roots t' of $H(X) = 0$ give a presentation of the new group as a subgroup of the old (see above) so do the arrangements corresponding to the roots t' of any other factor on the right side of (1) (because these factors are factors of $F(X)$ irreducible over K'). By the description of normal subgroups in §40, what is to be shown is that if t_1 is any root of $H(X, r)$ and if t' is any root of $G(X)$, then the substitution which carries $a = \phi_a(t)$ to $\phi_a(t')$, b to $\phi_b(t')$, c to $\phi_c(t'), \ldots$ carries the arrangement $\phi_a(t_1)$, $\phi_b(t_1)$, $\phi_c(t_1), \ldots$ to an arrangement of the form $\phi_a(t'_1)$, $\phi_b(t'_1)$, $\phi_c(t'_1), \ldots$ where t'_1 is a root of the same factor $H(X, \alpha^i r)$ of (1) that t' is. To this end, let $t_1 = \psi(t)$ be an expression of t_1 as a polynomial in t with coefficients in K. Then $H(\psi(X), r)$ is a polynomial in X with coefficients in K' of which t is a root (by the assumption on t_1). Therefore (Lemma I) it is divisible by $H(X, r)$. The lemma then implies that $H(\psi(X), \alpha^i r)$ is divisible by $H(X, \alpha^i r)$ for all i. In particular, if t'_1 is defined to be $\psi(t')$ then t'_1 and t' are roots of the same factor $H(X, \alpha^i r)$ of (1). Thus it will suffice to show that the substitution corresponding to

$t \to t'$ carries $\phi_a(t_1)$ to $\phi_a(t'_1)$, $\phi_b(t_1)$ to $\phi_b(t'_1)$, etc. But this follows immediately from the first corollary of §41, which shows that $t \to t'$ corresponds to an automorphism of the field $K(a, b, c, \ldots) = K(t)$ which leaves elements of K fixed; application of this automorphism to $\psi(t) = t_1$ carries it to $\psi(t') = t'_1$ and therefore carries $\phi_a(t_1)$, $\phi_b(t_1)$, $\ldots$ to $\phi_a(t'_1)$, $\phi_b(t'_1)$, $\ldots$, as was to be shown. $\qquad\square$

§45 The preceding two articles show that if $f(x) = 0$ is solvable by radicals then its Galois group G has a sequence of subgroups $G \supset G' \supset G'' \supset \cdots \supset G^{(v)}$ such that each group $G^{(i)}$ is a normal subgroup of prime index in its predecessor $G^{(i-1)}$, and such that the final subgroup $G^{(v)}$ consists of the identity substitution alone. Indeed, for this one needs only to take the sequence of field extensions $K \subset K' \subset K'' \subset \cdots \subset K^{(\mu)}$ implied by the solution by radicals, to take the Galois groups of the equation $f(x) = 0$ relative to each of the fields $K^{(i)}$, and to disregard groups that coincide with their predecessors.

A group G is said to be *solvable* if it has such a sequence of subgroups. Thus it has been shown that if an equation* $f(x) = 0$ is solvable by radicals then its Galois group is solvable. Galois showed that this necessary condition for solvability by radicals is also sufficient. More specifically, he showed that if G is the Galois group of $f(x) = 0$ over K and if $G \supset G' \supset \cdots \supset G^{(v)}$ is a sequence of subgroups of G in which each $G^{(i)}$ is a normal subgroup of prime index in its predecessor and $G^{(v)} = \{\text{identity}\}$ then there is a sequence of field extensions $K \subset K' \subset K'' \subset \cdots \subset K^{(\mu)}$ such that $K^{(\mu)}$ contains all the roots of $f(x) = 0$ and such that each extension $K^{(i-1)} \subset K^{(i)}$ is obtained by adjunction of a pth root of a quantity in $K^{(i-1)}$ for some prime p (depending on i). The main element in the proof of this fact is the proof of a converse of the proposition of §44. This converse is the subject of the next article.

§46 Proposition. *Let G be the Galois group of $f(x) = 0$ over K and let G' be a normal subgroup of G of prime index p. Assume also that K contains a primitive pth root of unity α, that is, a solution of $\alpha^p = 1$ with $\alpha \neq 1$. Then there is an element k of K such that if $K' \supset K$ is the field obtained by adjoining a pth root of k to K then the Galois group of $f(x) = 0$ over K' is G'.*

PROOF. The basic idea of the proof is the idea of the Lagrange resolvent, an element whose pth power is known. In fact, in his proof of this proposition, Galois uses an element of the form $\theta + \alpha\theta_1 + \alpha^2\theta_2 + \cdots + \alpha^{p-1}\theta_{p-1}$ which is reminiscent of the Lagrange resolvent or, perhaps more immediately, of Gauss's use of an analogous technique in the reduction of the finding of pth roots of unity to the extraction of roots. (See §25.)

---

* For reasons explained in §31, only equations $f(x) = 0$ with distinct roots need be considered.

The objective is to determine $\theta, \theta_1, \theta_2, \ldots, \theta_{p-1}$ in such a way that the Galois group merely permutes them cyclicly; then $\theta + \alpha\theta_1 + \alpha^2\theta_2 + \cdots + \alpha^{p-1}\theta_{p-1}$ is multiplied by a power of α by elements of the Galois group, and its pth power is invariant under the Galois group, so that, by Proposition I (§41), the pth power is a known quantity.

For his choice of θ, Galois simply says that it should be an element of $K(a, b, c, \ldots)$—"a function of the roots"—which is invariant under G' but not invariant under G. (However, at this crucial point—and one should think of Poisson here—Galois says the opposite of what he means and requires that θ "not vary for other substitutions" than those in G' instead of "does not remain invariant for other substitutions.") In a marginal note, he indicates the following proof that such θ's exist.

Let G be presented as in Galois' definition of the Galois group, with the rows corresponding to conjugates of the Galois resolvent t. Let $t, t', \ldots, t^{(\mu-1)}$ be the conjugates of t which correspond to rows in the presentation of the subgroup $G' \subset G$. (Then μ is the number of substitutions in G' and $p\mu$ the number in G.) The coefficients of $C(X) = (X - t)(X - t') \cdots (X - t^{(\mu-1)})$ are the elementary symmetric functions of $t, t', \ldots, t^{(\mu-1)}$ and are therefore invariant under the substitutions of G' because these substitutions merely reorder $t, t', \ldots, t^{(\mu-1)}$. If the coefficients of C were invariant under G then, by Proposition I, they would all be in K, and t would be a root of the polynomial $C(X)$ of degree μ with coefficients in K, contrary (by Lemma I of §41) to the fact that t is the root of an irreducible polynomial of degree μp with coefficients in K (because μp is the number of elements in the Galois group over K). Therefore, at least one coefficient of $C(X)$ is not invariant under G and has the properties required of θ.

Let θ be chosen and let S be a substitution of the Galois group which does not leave θ invariant, say $S\theta = \theta_1 \neq \theta$. Define θ_2 to be $S\theta_1$, θ_3 to be $S\theta_2$, and so forth. Galois says that "since p is a prime number, this sequence cannot end until the term θ_{p-1}, after which one has $\theta_p = \theta$, $\theta_{p+1} = \theta_1$, and so forth." However, he gives no proof. In order to prove that $\theta_i = \theta_{i+j}$ if and only if j is divisible by p, one can proceed as follows.

Think of the presentations of G as being arranged into p presentations of G'. If $a, b, c, \ldots$ is the arrangement in the 1st row of the first presentation of G' then $Sa, Sb, Sc, \ldots$ occurs in a *different* presentation of G' because S is not in G'. Then, because G' is by assumption a *normal* subgroup, the presentation of G' in which $Sa, Sb, Sc, \ldots$ occurs is obtained simply by applying S to the first presentation of G'. Thus, S acts* on the p presentations of G'. If P represents the first presentation of G' then $S(P), S^2(P), S^3(P), \ldots$ are

---

* More generally, if G' is a normal subgroup of index n in G then G acts on the n presentations of G' contained in a presentation of G; since two elements S and T of G act in the same way on these n presentations if and only if $S^{-1}T$ is in G', the *quotient group* G/G' acting on the n presentations gives a presentation of G/G'. The notion of a quotient group does not seem to be needed elsewhere in the book, so it has been circumvented here. The main idea in the proof of the present proposition is that G/G', being a group of prime order, must be cyclic.

presentations of G'. Let v be the least positive integer such that $S^v(P) = P$. Then S^v applied to an arrangement in P is another arrangement in P and S^v is in G'. Therefore, for any presentation P_1 of G', $S^v(P_1) = P_1$. If $1 \le j < v$ then $S^j(P_1) \ne P_1$ because $S^j(P_1) = P_1$ would imply $S^j(P) = P$, contrary to the definition of v. Therefore, the action of S divides the p presentations of G' into subsets containing v presentations each. Since $v \ge 2$ and p is a prime, it follows that $v = p$. This shows that S^p is in G' and therefore that $S^p\theta = \theta$. If $S^j\theta = \theta$ for $1 \le j < p$ then, since there are integers a and b such that $aj = bp + 1$ (that is, j is invertible mod p), it follows that $\theta = S^j\theta = S^{2j}\theta = \cdots = S^{aj}\theta = S^{bp+1}\theta = S\theta$, contrary to assumption. Therefore, as Galois stated, $\theta, \theta_1, \theta_2, \ldots, \theta_{p-1}$ are all distinct, but $\theta_p = \theta$, $\theta_{p+1} = \theta_1, \ldots$. Moreover, if T is any substitution in the Galois group, then $T(P) = S^j(P)$ for some j because $P, S(P), S^2(P), \ldots, S^{p-1}(P)$ covers all the presentations of G'. This shows that *any substitution T in the Galois group is equal to a substitution in G' (namely, $S^{-j}T$) followed by a power S^j of S.*

Now let $r = \theta + \alpha\theta_1 + \alpha^2\theta_2 + \cdots + \alpha^{p-1}\theta_{p-1}$. Galois' "proof" of the proposition is merely to state that r^p is invariant under the Galois group, and is therefore known, and that an extension K' of the desired type is obtained by adjoining a pth root of this known quantity. The steps here can be filled in as follows.

A crucial part of the proof, which Galois makes no mention of, is the proof that $r \ne 0$. For this, let $r_j = \theta + \alpha^j\theta_1 + \alpha^{2j}\theta_2 + \cdots + \alpha^{(p-1)j}\theta_{p-1}$, for $j = 1, 2, \ldots, p - 1$. Since α^j is a pth root of unity and $\alpha^j \ne 1$, r_j can be used in place of r in the arguments that follow. Therefore one can assume that $r \ne 0$ unless *all* the r's are zero, that is, $r_1 = r_2 = \cdots = r_{p-1} = 0$. But if all the r's were zero then

$$r_1 + r_2 + \cdots + r_{p-1} = (p - 1)\theta + (\alpha + \alpha^2 + \cdots + \alpha^{p-1})\theta_1$$
$$+ (\alpha^2 + \alpha^4 + \cdots + \alpha^{2p-2})\theta_2 + \cdots$$
$$= p\theta - \theta - \theta_1 - \theta_2 - \cdots$$

would also be zero, that is, $p\theta = \theta + \theta_1 + \cdots + \theta_{p-1}$; this would give $S\theta = \frac{1}{p}[S\theta + S\theta_1 + \cdots + S\theta_{p-1}] = \frac{1}{p}[\theta_1 + \theta_2 + \cdots + \theta_{p-1} + \theta] = \theta$, contrary to assumption. Thus $r \ne 0$ can be assumed. If T is any element of the Galois group then, as was noted above, $T = S^jU$ where U is in G' and j is an integer, $0 \le j < p$. Now $U\theta = \theta$ so $U\theta_1 = US\theta = SS^{-1}US\theta = S\theta = \theta_1$, because $S^{-1}US$ is in G' by virtue of the fact that G' is a normal subgroup of G. Similarly, $U\theta_2 = \theta_2, \ldots, U\theta_{p-1} = \theta_{p-1}$. Therefore $Ur = r$ and $Tr = S^jr = \theta_j + \alpha\theta_{j+1} + \cdots + \alpha^{p-1}\theta_{j+p-1} = \alpha^{p-j}r$, from which it follows that $T(r^p) = (Tr)^p = r^p$. Thus r^p is invariant under the Galois group and by Proposition I is an element of K, say $r^p = k$.

Now let K' be the subfield $K(r)$ of $K(a, b, c, \ldots)$, that is, all elements of $K(a, b, c, \ldots)$ that can be expressed as polynomials in r with coefficients in K. By the Proposition of §44, the Galois group of $f(x) = 0$ over K' is either G or a normal subgroup of G of index p. It cannot be G because $Sr = \alpha^{-1}r \ne r$ (because $r \ne 0$ and $\alpha^{-1} \ne 1$), which shows that S is an element of G which does not leave all elements of K' fixed. Thus the Galois group of

$f(x) = 0$ over K' is a normal subgroup G'' of G of index p. If T is in G'' then $Tr = r$, and, as above, $T = S^j U$ where $U \in G'$. Since $Tr = S^j Ur = S^j r = \alpha^{p-j} r$, and $r \neq 0$, the equation $Tr = r$ implies $\alpha^{p-j} = 1, \alpha^j = \alpha^p = 1, p$ divides j, $S^j =$ identity, $T = U$, and therefore T is in G'. Thus $G'' \subset G'$. Since both subgroups have index p in G, it follows that $G'' = G'$, and the proof of the proposition is complete.   □

§47 Theorem. *Let $f(x) = 0$ be an equation with distinct roots whose Galois group G over the field K is solvable, that is, has a sequence of subgroups $G \supset G_1 \supset G_2 \supset \cdots \supset G_v$ in which each G_i is a normal subgroup of prime index in its predecessor and G_v contains the identity substitution alone. Then $f(x) = 0$ can be solved by radicals, that is, one can construct a sequence of field extensions $K \subset K' \subset K'' \subset \cdots \subset K^{(\mu)}$ such that $f(x) = 0$ has $n = \deg f$ roots in $K^{(\mu)}$ and such that the elements of any $K^{(i)}$ can be expressed rationally in terms of elements of its predecessor and a fixed radical of a fixed element of its predecessor—in fact the elements can be expressed as polynomials in the radical with coefficients in $K^{(i-1)}$.*

PROOF. This follows immediately from the proposition of the preceding article, except for the problem of assuming the existence of roots of unity in K. Let a sequence of subgroups of G be given as in the statement of the theorem, and let $p_1, p_2, \ldots, p_v$ be the primes which occur as indices of subgroups in the sequence. Then, as Gauss showed, the p_1th roots of unity can be obtained by adjoining a succession of radicals to the rational field $\mathbb{Q}$ and *a fortiori* by adjoining radicals to K. (See note in §43.) Repetition of this process constructs a field $K^{(\lambda)} \supset K^{(\lambda-1)} \supset \cdots \supset K' \supset K$ which contains all p_1th, p_2th, $\ldots$, p_vth roots of unity. Let $\tilde{G}$ be the Galois group of $f(x) = 0$ over $K^{(\lambda)}$. Then $\tilde{G}$ is a subgroup of G and the proposition of the preceding article will easily prove the theorem if it is shown that $\tilde{G} \supset \tilde{G} \cap G_1 \supset \tilde{G} \cap G_2 \supset \cdots \supset \tilde{G} \cap G_v = \{$identity$\}$ is* a sequence of subgroups of $\tilde{G}$ in which each subgroup is a normal subgroup of prime index in its predecessor and the indices which occur are all included among the prime indices $p_1, p_2, \ldots, p_v$. Thus, the theorem will follow if it is shown that:

Lemma. *If G_i is a normal subgroup of G_{i-1} with prime index p and if $\tilde{G}$ is any subgroup of a group containing G_{i-1} then either $G_i \cap \tilde{G}$ is equal to $G_{i-1} \cap \tilde{G}$ or it is a normal subgroup of index p.*

PROOF. As was seen in the proof of §46, every element of G_{i-1} permutes the p presentations of G_i contained in a presentation of G_{i-1}. Therefore every element of $G_{i-1} \cap \tilde{G}$ permutes these presentations of G_i. Moreover, an element of G_{i-1} acts as the identity permutation of them if and only if it

---

* More precisely, it becomes such a sequence when groups which coincide with their predecessors are omitted.

lies in G_i. Thus $G_{i-1} \cap \tilde{G} = G_i \cap \tilde{G}$ if and only if no element of $G_{i-1} \cap \tilde{G}$ moves a presentation of G_i. Otherwise some element S of $G_{i-1} \cap \tilde{G}$ moves a presentation of G_i and, as was seen in §46, S permutes the p presentations cyclicly. In this case, the p images of a presentation of $G_i \cap \tilde{G}$ under powers of S comprise a presentation of $G_{i-1} \cap \tilde{G}$, which shows that $G_i \cap \tilde{G}$ is a normal subgroup of index p. □

§48 This ends the statement of the basic facts of Galois theory. In brief, the theory associates to a given equation $f(x) = 0$, over a given field K, a finite group called its *Galois group*, and shows that the equation can be solved by radicals if and only if its Galois group is solvable. Galois used the theory to derive the known solutions of equations by radicals, to show that the general equation of degree ≥ 5 cannot be solved by radicals, and to prove a theorem about the solution of irreducible equations of prime degree by radicals. It is clear that he saw these as simply the first applications which would be the most accessible and striking ones for his contemporaries. Unfortunately for mathematics, he did not live to develop the more profound applications and extensions of his theory. Even today, mathematics is probably the poorer as a result of this tragedy.

The following thirteen articles (§§49–61) are devoted to giving a firmer foundation to Galois theory by spelling out more carefully what is meant by "the roots $a, b, c, \ldots$" of a given equation. This will give a clearer meaning to the field $K(a, b, c, \ldots)$ and therefore to the Galois resolvent. Following §61, Galois theory will be briefly reviewed and Galois' applications, along with some others, will be given.

Sixth Exercise Set

1. Show that the six-element group of all substitutions of three letters is solvable by finding a sequence of subgroups with the required properties. Use this sequence of subgroups and the method of the text to give a method of solving cubic equations. Be as explicit as possible.

2. Do the same for the twenty-four element group of all substitutions of four letters.

3. Let a, b, c, d be the roots of a quartic equation and let $P = ab + cd$, $Q = ac + bd$, $R = ad + bc$. Show that the discriminant of the cubic of which P, Q, R are the roots is equal to the discriminant of the quartic of which a, b, c, d are the roots. Use this fact to find the discriminant of the quartic. [Assume that $a + b + c + d = 0$ to simplify the computations. Describe how the formula in this special case can be used to find the general formula without carrying this out.]

4. Show that if one root of an irreducible equation can be expressed in terms of radicals then all roots can. [Let K, K' be as in the proposition of §44 and let f be a polynomial with coefficients in K which is irreducible over K. Show that if $f = g_1 g_2 \cdots g_k$ is a factorization of f into factors irreducible over K' and if all roots of g_1 can be expressed in terms of radicals then all roots of f can be expressed in terms of radicals.]

5. Prove as a corollary to the proposition of §46 that for any prime p the pth roots of unity can be obtained by adjoining a succession of radicals to the rational field $\mathbb{Q}$.

6. Galois changed his statement of Proposition II, possibly on the night before his duel, eliminating the assumption that the degree of the auxiliary equation be prime. In all likelihood he intended to state:

If a root r of an irreducible polynomial of degree m is adjoined to K, then the index of the new Galois group (that of $f = 0$ over $K(r)$) *in the old Galois group* (that of $f = 0$ over K) *is a divisor of m.*

Prove this. [A slight modification of the proof in the text.] Historical note: Galois' actual statement that the old group is partitioned into "groups" shows that he was not using the word "group" in the modern sense, because it is not the substitutions, but the arrangements, that are partitioned. Each "group" in the partition is a presentation of a subgroup of the old group; these subgroups are *conjugates* (not necessarily distinct) of the new group.

7. Prove Galois' Proposition III (also added to his treatise at a later time, possibly on the night before his duel):

If, in Proposition II, all roots of the auxiliary equations are adjoined to K then the new group is a normal subgroup of the old.

8. Liouville's proof of Proposition II did not follow Galois' suggestion of a proof. He let $G(X)$, the irreducible polynomial over K of which t is a root, be decomposed $G(X) = H_1(X)H_2(X)\ldots H_\mu(X)$ as a product of irreducible polynomials with coefficients in $K(r)$. For a polynomial $g(X)$ with coefficients in $K(r)$, let the *norm* $Ng(X)$ be the polynomial with coefficients in K obtained by replacing r in $g(X)$ by each of the m values r, r', $r'', \ldots, r^{(m-1)}$, multiplying these m polynomials, and replacing symmetric polynomials in the r's by their values in K. Then $G^m = NG = NH_1 \cdot NH_2 \cdots NH_\mu$. Each NH_i is therefore a power of G. This power is the same in all cases because the H's have equal degree. Thus $m = \nu\mu$, where ν satisfies $NH_1 = G^\nu$. Thus $\mu = \deg G/\deg H$ divides m, as was to be shown. Fill in the steps of this proof.

9. Prove Galois' Proposition IV: If u is in $K(a, b, c, \ldots)$ — that is, if u is a polynomial in the roots $a, b, c, \ldots$ of $f(x) = 0$ — then the Galois group of $f(x) = 0$ over $K(u)$ is equal to the subgroup of the Galois group of $f(x) = 0$ over K consisting of those substitutions which leave u fixed.

10. Prove Dedekind's "reciprocity theorem" of Galois theory (1855, first published in 1982, [S1]): Let f and g be irreducible polynomials with coefficients in K. Let f decompose into irreducible factors $f = f_1 f_2 \ldots f_\mu$ over $K(b)$ where b is a root of g. Similarly, let g decompose $g = g_1 g_2 \ldots g_\nu$ over $K(a)$ where a is a root of f. Then $\mu = \nu$ and the g_i can be reordered in such a way that the ratio of $\deg f_k$ to $\deg g_k$ is the same for all $k = 1, 2, \ldots, \mu$. [Show that a_i and a_j are roots of the same f_k if and only if there is an S in the Galois group of $fg = 0$ over K with $Sa_i = a_j$ and $Sb_1 = b_1$, where $b = b_1$ is a fixed root of g. Fix a root a_1 of f and define g_k to be the factor of g over $K(a_1)$ of which $S^{-1}b_1$ is a root, where S is in the Galois group of $fg = 0$ over K and Sa_1 is a root of f_k. Show that $\deg f_k/\deg f = \deg g_k/\deg g$ by showing both are the proportion of S's carrying a_1 to a root of f_k (or b_1 to a root of g_k).]

11. Deduce from Dedekind's theorem of Exercise 10 that if f is a polynomial with coefficients in K and if K' is obtained from K by adjoining *all* roots of a polynomial g with coefficients in K then all irreducible factors of f over K' have the same degree.

12. Relate the solution of the cubic in Exercise 1 to the solution in §5.

13. Prove that if K' is an extension of K then the Galois group of $f(x) = 0$ over K' coincides with its Galois group over K if and only if $K(a, b, c, ...) \cap K' = K$. (Both $K(a, b, c, ...)$ and K' are contained in $K'(a, b, c, ...)$.) In other words, the Galois group is reduced if and only if one of the newly adjoined quantities in K' can be expressed rationally in terms of the roots of f and previously known quantities.

Roots and Splitting Fields

§49 Galois and his predecessors talked about "the roots" $a, b, c, ...$ of an equation without saying what these roots were or justifying the assumption that roots exist. Questions of mathematical existence are among the most profound questions in the philosophy of mathematics, and it would be difficult if not impossible to guess how Galois, Lagrange, Vandermonde, Newton, Girard, and the other founders of modern algebra might have answered these questions had they dealt with them. The answers that are given below are essentially the ones given several decades after Galois' work by Kronecker, who was not only a great mathematician but also an important philosopher of mathematics.

The central assumption of Galois theory, which Galois did not justify and which was not justified in the exposition of Galois theory above, is the assumption that it is meaningful to talk about the field $K(a, b, c, ...)$ of all "functions of the roots" $a, b, c, ...$ of the given equation $f(x) = 0$. In modern terminology, such a field is a special case of what is called a *splitting field* for $f(x)$.

Definition. Let $f(x)$ be a polynomial with coefficients in a field K. A *splitting field* for $f(x)$ is a field L which contains K and which has the property that $f(x)$ can be written as a product of polynomials of the first degree with coefficients in L.

If any splitting field L of $f(x)$ over K is known, then Galois' constructions of the Galois resolvent and the Galois group can be carried out within L. If, say, $f(x) = (\alpha_1 x + \beta_1)(\alpha_2 x + \beta_2) \cdots (\alpha_n x + \beta_n)$ where α_i and β_i are in L, then one can set $a = -\beta_1/\alpha_1$, $b = -\beta_2/\alpha_2$, $c = -\beta_3/\alpha_3, ...$ to find $f(x) = \alpha_1 \alpha_2 ... \alpha_n (x - a)(x - b)(x - c) ....$ Assuming, as always, that $a, b, c, ...$ are distinct (that is, $f'(x)$ has no factor in common with $f(x)$), one can find (see §32) a Galois resolvent $t = Aa + Bb + Cc + \cdots$ (A, B, C, ... integers) *as an element of* L. Then $K(a, b, c, ...) = K(t)$ is a subfield of L and all the constructions of Galois theory are meaningful in terms of L.

To establish a firm foundation for Galois theory, it will suffice, therefore, to prove the existence of a splitting field for any given polynomial. This is a much more satisfactory way of stating the problem than to say that in some sense it is meaningful to speak of "the roots" of an equation. What is to be shown is that there is a mathematical structure within which computations can be carried out and within which the given equation has roots.

§50   The so-called fundamental theorem of algebra proves the existence of a splitting field for any equation $f(x) = 0$ with *rational* coefficients or even with coefficients from the field of complex numbers. This theorem states that *a polynomial equation* $Ax^n + Bx^{n-1} + \cdots + C = 0$, *in which the coefficients* $A, B, \ldots, C$ *are complex numbers,* $n > 0$, *and* $A \neq 0$, *always has a complex root.* Gauss devoted much effort to giving a rigorous proof of this theorem. Given that the equation $Ax^n + Bx^{n-1} + \cdots + C = 0$ has a root, say a, one can then divide by $(x - a)$ to obtain a polynomial of lower degree with complex coefficients and the process can be repeated to write $Ax^n + Bx^{n-1} + \cdots$ as a product of linear factors $A(x - a)(x - b) \ldots$. Thus *the field of complex numbers is a splitting field for any polynomial with complex coefficients.*

There are two major reasons why this theorem is not a suitable foundation for Galois theory (and therefore not a suitable "fundamental theorem" of algebra). The first is that Galois theory applies to equations that do *not* have complex coefficients. In fact, the first objective of Galois theory was to prove that the *general* quintic equation $x^5 + Ax^4 + Bx^3 + Cx^2 + Dx + E = 0$, in which the coefficients are *variables* (see the Preface), is not solvable by radicals in the way that the general quadratic, cubic, and quartic equations are.

The second reason is that it is not a theorem of algebra at all. The complex numbers, like the real numbers, are constructed by *transcendental* means, involving limits or Dedekind cuts in some essential way. Accordingly, the proof of the theorem involves topological considerations, limits, and other nonalgebraic ideas, and the roots it provides are described in a transcendental rather than an algebraic way.

Of course, if there were no alternative, then these objections could both be overruled as being aesthetic rather than substantial. However, there is a natural approach to the foundations of the subject which overcomes both objections very easily.

Construction of a Splitting Field

§51   The theorem on simple algebraic extensions (§34) constructs a field containing a root of a given *irreducible* polynomial. It can be used to construct a splitting field of a given polynomial f as follows.

If the given polynomial f with coefficients in a given field K is irreducible then the theorem of §34 gives a field $K(a)$ containing K and containing a root a of $f(x) = 0$. Otherwise f is reducible, say $f = f_1 f_2$ where f_1 and f_2 have coefficients in K and have degree less than $\deg f$. If either f_1 or f_2 is irreducible, say f_1 is, then the theorem on simple algebraic extensions constructs a field $K(a)$ containing K and a root a of $f_1(x) = 0$. Since $f = f_1 f_2$, a is a root of f, and this gives an extension of K which contains one root (at least) of f. If neither f_1 nor f_2 is irreducible then $f_1 = f_3 f_4$, $f_2 = f_5 f_6$ where f_3 and f_4 have degree less than $\deg f_1$, f_5 and f_6 degree less than $\deg f_2$. If any one of the factors $f = f_3 f_4 f_5 f_6$ is irreducible then, as before, the theorem on simple algebraic extensions can be used to construct a field $K(a) \supset K$ containing one root (at least) of f. Since each factorization reduces the degree, this process must eventually lead to an irreducible factor of f and therefore to an extension field $K(a) \supset K$ containing a root of f.

Let $K(a) \supset K$ be a field containing a root a of f. Then by the Remainder Theorem (§37) $f(x) = (x - a)\tilde{f}(x)$ where $\tilde{f}$ has coefficients in $K(a)$ and degree one less than $\deg f$. The entire process can now be repeated to give a field $K(a, b)$ which contains $K(a)$ and one root (at least) of $\tilde{f}$. Then $f(x) = (x - a)(x - b)\tilde{\tilde{f}}(x)$, where $\tilde{\tilde{f}}$ has coefficients in $K(a, b)$. Repetition of this process $\deg f$ times gives a field $K(a, b, c, \ldots)$ in which f has $\deg f$ roots and splits into linear factors. Thus $K(a, b, c, \ldots)$ is a splitting field of f, as desired.

The Need for a Factorization Method

§52 The argument of the preceding article raises a fundamental question in the philosophy of mathematics. Is it valid to say that a given polynomial f is either factorable or irreducible? On its face, the search for a factorization of a polynomial involves *infinitely many* trial factors. Is it valid to imagine that these trials can *all* be carried out to determine whether the polynomial is reducible?

For the last 80 years or so, largely as a result of the influence of Cantor and Hilbert, such arguments have been widely accepted among mathematicians, despite the objections of Brouwer and his followers in the intuitionist school. In my opinion, the objections are entirely correct. The argument of the preceding article does not prove the existence of a field with the required property. Rather, it is an indication of how one *might* construct such a field if one were able to factor polynomials or prove they are irreducible whenever it is called for.

What is at issue is the nature of mathematical "existence." The Cantorian school believes in some form of existence in an unseen world of Platonic ideals. To me, such existence has little, if any, meaning. The only satisfactory proof that a field exists is one that shows explicitly how computations in the field are to be performed—how to represent elements of the field, what it means to say two elements are equal, how to multiply elements, and so forth. In other words, the only satisfactory proof is a *construction* of the field.

Whether or not one believes that something can be proved to exist without having been constructed, almost everyone will agree that a constructive proof is to some extent preferable and will acknowledge that there is some value in making the effort to give a constructive proof of the existence of a splitting field. For me, this effort is necessary to put Galois theory on a firm foundation.

§53 As the argument of §51 shows, in order to prove the existence of a splitting field for a polynomial with coefficients in a field K, it suffices to prove that there is a *method of factoring* polynomials with coefficients in K and a method of factoring polynomials with coefficients in each of the fields $K(a)$, $K(a, b), \ldots$ that occur in the construction. Here a method of factoring polynomials is a procedure which can be applied to any given polynomial and will arrive after a finite number of steps at either a factorization of the polynomial or a proof that it is irreducible. The remainder of this section (ending with §61) is devoted to the factorization of polynomials.

Of course the method of factorization depends on the coefficient field K, and it would be unreasonable to expect to be able to factor polynomials over *all* fields, because the notion of a field is such a general one. For what fields K should one attempt to find factorization methods? Since Galois theory must at the very least apply to polynomials with rational coefficients, it must at the very least be shown that *there is a factorization method for the coefficient field $\mathbb{Q}$ of rational numbers.* This alone is not enough to establish Galois theory for polynomials with rational coefficients because, even when $K = \mathbb{Q}$, the second stage of the construction of §51 calls for a factorization over $\mathbb{Q}(a)$, the third stage for a factorization over $\mathbb{Q}(a, b)$, and so forth. In order to prove in one stroke that all these factorizations are possible, it is natural to try to prove that *given a factorization method for the coefficient field K, one can find a factorization method for the coefficient field $K(a)$ obtained by adjoining to K a root a of an irreducible polynomial with coefficients in K.* This will be proved below, and it will follow from this and the preceding theorem that there is a splitting field for any polynomial with rational coefficients. Finally, Galois himself clearly thought of his theory as applying to equations with *literal* as well as *numerical* coefficients, so to put this part of the theory on a firm foundation it is necessary to study coefficient fields which contain indeterminates. The relevant construction here is the adjunction of an indeterminate to a given field K. If a denotes the indeterminate, this gives the field $K(a) = \{p/q: p$ and q are polynomials in a with coefficients in K and $q \neq 0\}$, in which the operations of arithmetic are carried out in the obvious way and in which p_1/q_1 and p_2/q_2 are regarded as being *equal* if $p_1 q_2 = p_2 q_1$. Such an extension $K(a)$ is called a *transcendental* extension of K, as opposed to the adjunction of a root of an irreducible polynomial, which is called an *algebraic* extension of K. Fields containing several "indeterminates" can be obtained by adjoining the indeterminates serially in this way. Factorization over such fields is made possible by the theorem that *given a factorization method for the coefficient field K, one can find a factorization method for the coefficient field K(a) obtained by adjoining to K an indeterminate a.* These three theorems combine to prove that there is a factorization method for any coefficient field K that is obtained from the rational number field ℚ by a finite number of algebraic and/or transcendental adjunctions. Therefore, by the argument of §51, there is a splitting field for any polynomial with coefficients in any such field.

Unique Factorization into Irreducibles

§54 It will be useful to begin the investigation of the factorization of polynomials with coefficients in a field K with the observation that *if* a particular polynomial can be written as a product of irreducible polynomials then this can be done in essentially *only one* way. The qualification "essentially only one" is necessary here in order to take account of the fact that if c is any nonzero element of K then one factor can be multiplied by c and another can be multiplied by $1/c$, which changes the factorization in a superficial way. The statement about unique factorization can be made precise as follows.

A nonzero polynomial of degree 0—that is, a nonzero element of K regarded as a polynomial—is called a *unit*. A polynomial* f is called *irreducible* if it is not a unit and if the only factorizations of it (into factors with coefficients in K) are the trivial ones in which one factor is a unit (and, consequently, the other is the inverse of this unit times f). The theorem of unique factorization then states that if $f_1 f_2 \cdots f_\mu = g_1 g_2 \cdots g_\nu$ where the f's and g's are all irreducible polynomials then $\mu = \nu$ and the g's can be reordered in such a way that each g_i ($i = 1, 2, \ldots, \mu$) is a unit times the corresponding f_i.

The proof of this theorem is exactly the same as the proof of unique factorization for integers. As for integers, the key step is to prove:

Theorem. *Irreducible polynomials are prime, that is, if an irreducible polynomial f divides a product gh of two polynomials then it must divide one of the factors.*

PROOF. As for integers, the key step in the proof of this fact is the Euclidean algorithm (see §35). Let f, g, h be as in the statement of the theorem. The Euclidean algorithm applied to f and g gives a polynomial of the form $d = af + bg$ where a and b are polynomials and where d divides both f and g. Since f is irreducible and d divides f, either d is a unit or it is a unit times f. If it is a unit times f then, since d divides g, it follows that f divides g

---

* According to this definition, 0 is not irreducible. In the study of factorization, 0 plays no role, and therefore it is irrelevant whether one chooses to call it irreducible.

and f divides one of the factors of gh, as was to be shown. If d is a unit, then $dh = afh + bgh$ shows that f divides dh (f divides gh by assumption) and therefore that f divides h. Thus f divides either g or h, as was to be shown. $\square$

The deduction of unique factorization from this theorem will be left to the reader (Exercise 1).

Factorization Over $\mathbb{Q}$

§55 It will be shown below (§§56 and 57) that a method for factoring polynomials with rational coefficients can be obtained by first finding a method for factoring polynomials with *integer* coefficients and by using this in the factorization of polynomials with rational coefficients. Consider, therefore, the factorization of polynomials with integer coefficients. In this case the *units*, the elements which divide 1 and therefore divide everything, are just the polynomials 1 and -1. A polynomial is *irreducible* if it is not a unit and if the only factorizations of it (into factors with integer coefficients) are the trivial ones in which one of the factors is a unit. The first objective is to give a method of factoring polynomials with integer coefficients, that is, a method which can be applied to a (nonzero) polynomial with integer coefficients to produce, in a finite number of steps, either a factorization of the polynomial or a proof that it is irreducible. The following simple method is due to Kronecker.

The factorization of polynomials of degree 0 is ordinary factorization of integers, which can be accomplished in a finite number of steps by simple trial and error. Therefore, let $f(x)$ be a polynomial of degree $n > 0$ with coefficients in $\mathbb{Z}$. If f has a nontrivial factorization $f(x) = g(x)h(x)$ over $\mathbb{Z}$ then either g or h must have degree $\leq n/2$. Therefore it will suffice to give a procedure which either gives a nontrivial factor of $f(x)$ of degree $\leq n/2$ or shows that there is no such factor.

Kronecker's method is based on the fact that a polynomial of degree m is completely determined once $m + 1$ of its values are known. For example, if g_1 and g_2 are polynomials of degree $\leq m$ and if $g_1(0) = g_2(0)$, $g_1(1) = g_2(1), \ldots, g_1(m) = g_2(m)$ then g_1 and g_2 are identical, because $g_1 - g_2$ is then a polynomial of degree $\leq m$ that is divisible by $x(x - 1)(x - 2)\cdots$ $(x - m)$, and only the zero polynomial satisfies these conditions.* Moreover, the Lagrange interpolation formula

$$g(0)\frac{(x - 1)(x - 2)\ldots(x - m)}{(-1)(-2)\ldots(-m)} + g(1)\frac{x(x - 2)(x - 3)\ldots(x - m)}{1(1 - 2)(1 - 3)\ldots(1 - m)} + \cdots$$

$$(1)$$

gives an explicit formula for a polynomial with *rational* coefficients which

---

* Note the parallel with the argument at the end of §11.

has degree at most m and agrees with g at $0, 1, 2, \ldots, m$ and is therefore identical with g. Thus, if the values of g at $0, 1, \ldots, m$ are known, g is known.

Now if $f(x) = g(x)h(x)$ where f, g, h have integer coefficients then, for each integer j, $g(j)$ divides $f(j)$. If $f(j) = 0$ for any j then $f(x)$ is divisible by $x - j$ and this is a nontrivial factorization of $f(x)$ (into factors with integer coefficients) unless $f(x) = \pm(x - j)$, in which case $f(x)$ is irreducible. Therefore it can be assumed that $f(j) \neq 0$ for $0 \leq j \leq n/2$. Then, for each j, $f(j)$ has a finite number of divisors (because it has no divisors with absolute value greater than $|f(j)|$). Thus there is a finite set of possibilities for the values of $g(0), g(1), \ldots, g(m)$ where m is the greatest integer $\leq n/2$ and $g(x)$ is a polynomial of degree $\leq m$ that divides $f(x)$. (Explicitly, there are $\mu_0 \mu_1 \cdots \mu_m$ sets of possible values where μ_j is the number of divisors of $f(j)$.) Let each of these sets of values be listed, and with each of them list the corresponding polynomial (1). Strike from the list all entries in which (1) does not have integer coefficients, all entries in which division of $f(x)$ by (1) leaves a remainder or gives a quotient whose coefficients are not integers, and all entries in which (1) is ± 1. If any entry survives, it gives a nontrivial factorization of $f(x)$. Otherwise $f(x)$ is irreducible over ℤ. This completes the factorization of polynomials over ℤ.

§56 Now let $f(x)$ be a nonzero polynomial with *rational* coefficients and degree > 0. Let d be a nonzero integer divisible by the denominators of all of the coefficients of $f(x)$ (for example, let d be the product of all the denominators). Then $d \cdot f(x)$ is a polynomial with integer coefficients, say $d \cdot f(x) = F(x)$. Regarded as a polynomial with *rational* coefficients, $F(x)$ is a unit times $f(x)$; therefore, a factorization of $f(x)$ into irreducible factors implies one of $F(x)$ and conversely. Regarded as a polynomial with *integer* coefficients, $F(x)$ can be factored $F(x) = F_1(x)F_2(x) \ldots F_\mu(x)$ by the method of the preceding article into factors $F_i(x)$ that are irreducible as polynomials with integer coefficients. This method gives a factorization of $F(x)$ into polynomials irreducible over ℚ in the sense defined above because *a polynomial with integer coefficients which is irreducible over ℤ* (that is, a polynomial which is not ± 1 and has no factorization into polynomials with integer coefficients other than the trivial ones in which one of the factors is ± 1) *when regarded as a polynomial with rational coefficients is either a unit* (if its degree is 0) *or is irreducible* (if its degree is > 0). This important fact is proved as Corollary 1 to the theorem of the next article.

§57 Theorem. *Let F, G, and H be polynomials with integer coefficients such that F is irreducible over ℤ and F divides GH. Then either F divides G or F divides H. (Here "divides" means that the quotient has integer coefficients.) In other words, in the ring of polynomials with integer coefficients, irreducible elements are prime. (See the analogous theorem in §54.)*

Corollary 1. *If $F(x)$ is a polynomial with integer coefficients that has degree > 0 and is irreducible over ℤ, then $F(x)$ is irreducible over ℚ.*

The deduction of Corollary 1 from the theorem is Exercise 2. In the proof of the theorem that is given below, Corollary 1 is proved directly.

Corollary 2. *A representation of a polynomial with integer coefficients as a product of irreducibles is unique up to the order of the factors and their signs. That is, if $F_1 F_2 \ldots F_\mu = G_1 G_2 \ldots G_\nu$ and the F's and G's are all irreducible then $\mu = \nu$ and it is possible to reorder the G's so that $F_i = \pm G_i (i = 1, 2, \ldots, \mu)$.*

The deduction of this corollary is left as an exercise. It is even easier than the analogous statement in §54. (See Exercise 1.)

PROOF OF THE THEOREM. Consider first the case deg $F = 0$, that is, the case where F is a prime integer p. This case of the theorem is known as *Gauss's lemma*.* Let p be a prime integer which divides GH. It is to be shown that p divides G or H. If $G(x) = a_\mu x^\mu + a_{\mu-1} x^{\mu-1} + \cdots + a_0$ is not divisible by p then there is a largest index i such that p does not divide a_i. Similarly, if $H(x) = b_\nu x^\nu + b_{\nu-1} x^{\nu-1} + \cdots + b_0$ is not divisible by p then there is a largest index j such that p does not divide b_j. The coefficient of x^{i+j} in GH is $\cdots + a_{i+1} b_{j-1} + a_i b_j + a_{i-1} b_{j+1} + \cdots$ and this coefficient is not divisible by p because $a_i b_j$ is not divisible by p (p is prime) but the terms preceding it are divisible by p (because $a_{i+1}, a_{i+2}, \ldots$ are) and the terms after it are also divisible by p (because $b_{j+1}, b_{j+2}, \ldots$ are). (In short, if i is the degree of G mod p and j is the degree of H mod p then $i + j$ is the degree of GH mod p.) Thus, if neither G nor H is divisible by p then GH is not divisible by p. The contrapositive of this statement is Gauss's lemma.

The next step in the proof of the theorem is to prove Corollary 1. For this, assume $F(x)$ is irreducible over the integers and assume $F(x) = g(x)h(x)$ where g and h are polynomials with rational coefficients. Let this equation be multiplied by a suitably chosen positive integer j to put it in the form $jF = G_1(x)H_1(x)$ where G_1 and H_1 have integer coefficients and the same degrees as g and h respectively. If p is any prime factor of j then, by Gauss's lemma, p divides either G_1 or H_1 and the equation can be divided by p to give $j'F(x) = G_2(x)H_2(x)$ where $j' = j/p$ and G_2 and H_2 have integer coefficients. Similarly, any prime factor of j' can be divided out and the process can be continued until one arrives at an equation of the form $F = G_3 H_3$.

---

* The name "Gauss's lemma" is used for various closely related propositions. The one which Gauss actually stated (Art. 42 of *Disquisitiones Arithmeticae*) was: *If g and h are polynomials in one variable with rational coefficients and with leading coefficients 1, and if gh has integer coefficients, then g and h both have integer coefficients.* This can be deduced from the case deg $F = 0$ of the theorem as follows. Let c and d be the least common denominators of the coefficients of g and h respectively, that is, the least positive integers such that $cg = G$ and $dh = H$ have integer coefficients. If p is a prime that divides cd, then p divides $GH = cdgh$ because gh has integer coefficients. By the theorem, p divides G or H; say it divides G. Since the leading coefficient of $G = cg$ is c, it follows that c/p is an integer. Since $G/p = (c/p)g$ has integer coefficients, this contradicts the definition of c. Similarly p cannot divide H, so there is no p, and $cd = 1$. Thus $c = d = 1$ and $g = G$, $h = H$ have integer coefficients, as was to be shown.

Since F is irreducible over $\mathbb{Z}$, either $G_3 = \pm 1$ or $H_3 = \pm 1$. Since G_3 has the same degree as g and H_3 the same degree as h, this shows that either g or h is a unit. Hence F is irreducible over $\mathbb{Q}$.

Thus, by the theorem of §54, the equation $F = GH$ implies that F divides G or H when they are considered as polynomials with rational coefficients, say $G = FQ$. It is to be shown that the quotient Q has integer coefficients. Again, multiplication by a suitable positive integer puts this equation in the form $jG = FQ_1$ where Q_1 has integer coefficients, and again the prime factors of j can be divided out one by one. Since F is irreducible over $\mathbb{Z}$ and $\deg F > 0$, F is not divisible by any integer. Therefore, these divisions leave F unchanged and lead to an equation $G = FQ_2$ where Q_2 has integer coefficients, as desired. $\square$

Factorization Over Transcendental Extensions

§58 This completes the proof that there is a factorization method for the coefficient field $\mathbb{Q}$. It remains to show that if there is a factorization method for the coefficient field K then there is also one for the coefficient field $K(a)$ where a is either algebraic or transcendental over K. The case where a is transcendental over K is easier and is used in the proof of the case where a is algebraic, so the transcendental case will be considered first.

Let $f(x) = A_\mu x^\mu + A_{\mu-1} x^{\mu-1} + \cdots + A_0$ be a polynomial with co-efficients A_i in $K(a)$ where a is transcendental over K and where K is a coefficient field for which there is a method of factorization. By definition, the A's are quotients of polynomials in a with coefficients in K. This equation can be multiplied by a suitable nonzero polynomial $q(a)$ (for example, the product of the denominators of the A's) to give $q(a)f(x) = F(a, x)$, where $F(a, x)$ is a polynomial in x whose coefficients are polynomials in a with coefficients in K—or, what is the same, where $F(a, x)$ is a *polynomial in two variables* a and x with coefficients in K. Since $q(a)$ is a unit of $K(a)$, the factor-ization of $f(x)$ over $K(a)$ is therefore equivalent to the factorization of $F(a, x)$ over $K(a)$. This can be accomplished in two steps:

(1) Show that $F(a, x)$ can be factored $F(a, x) = F_1(a, x)F_2(a, x) \ldots F_\nu(a, x)$ where each $F_i(a, x)$ is a polynomial in two variables which is irreducible in the sense that the only factorizations $F_i(a, x) = g(a, x)h(a, x)$ of F_i as a product of two polynomials in two variables are the trivial ones in which one of the factors is a nonzero element of K.
(2) Show that a polynomial in two variables which is irreducible in this sense is, when it is regarded as a polynomial in x with coefficients in $K(a)$, either a unit (if it does not contain x) or irreducible (if it does). Thus $f(x) = q(a)^{-1}F_1(a, x)F_2(a, x) \ldots F_\nu(a, x)$ expresses $f(x)$ as a unit times a product of irreducible polynomials in x with coefficients in $K(a)$, as required.

[Note the analogy with the method of factoring polynomials with co-efficients in $\mathbb{Q}$. There, $jf(x) = F(x)$ where $F(x)$ has coefficients in $\mathbb{Z}$. One shows that (1) $F(x) = F_1(x)F_2(x) \ldots F_\nu(x)$ where the $F_i(x)$ are polynomials that are irreducible as polynomials with *integer* coefficients and (2) such polynomials $F_i(x)$ are either units or are irreducible over $\mathbb{Q}$.]

To establish (1) is simple using a trick of Kronecker explained in the next article. The proof of (2) is almost exactly the same as the proof in the analogous case already considered, and will be left as an exercise (Exercise 3).

§59 The factorization of polynomials with coefficients in $K(a)$ (*a* transcendental over K) reduces, therefore, to the factorization of polynomials in two variables with coefficients in K. Kronecker gave the following trick for reducing the factorization of polynomials in two variables to the fac-torization of polynomials in one variable. Since, by assumption, the latter problem can be solved, this will provide a factorization method for poly-nomials with coefficients in $K(a)$.

Let $f(a, x)$ be a polynomial in the two variables a and x with coefficients in K. Let n be the degree of f in x. If $n = 0$ then f is a polynomial in a alone and can therefore be factored. Assume, therefore, that $n > 0$. Let N be any integer greater than n, and let $\tilde{f}(t) = f(t^N, t)$. If f can be factored, say $f = gh$, then $\tilde{f}$ can be factored $\tilde{f} = \tilde{g}\tilde{h}$ where $\tilde{g}(t) = g(t^N, t), \tilde{h}(t) = h(t^N, t)$. Since every integer ≥ 0 can be written in just one way in the form $iN + j$ where $i \geq 0$ and $0 \leqslant j < N$, and since g has degree $< N$ in x (because f does), it is easy to see (Exercise 5) that $\tilde{g}$ determines g uniquely. To factor f it will suffice, therefore, first to find all possible factors $\tilde{g}$ of $\tilde{f}$, next to find, for each of them, the corre-sponding g, and finally to check whether any of these g's divides f. All possible factors $\tilde{g}$ of $\tilde{f}$ (counting two factors as being the same if one is an element of K times the other) can be found by using the factorization method for poly-nomials in one variable to write $\tilde{f}$ as a product of irreducible factors $\tilde{f} = \phi_1\phi_2 \ldots \phi_\mu$ and letting $\tilde{g}$ range over all products of subsets of the ϕ's. The other two steps of the process are simple computations, and this completes the description of the factorization method.

Factorization Over Algebraic Extensions

§60 It remains to show that if polynomials can be factored over K then they can be factored over $K(a)$ where a is *algebraic* over K, say where $\phi(a) = 0$ where ϕ is a polynomial of degree $n > 1$, with coefficients in K, which is irreducible over K.

Let $f(x)$ be a polynomial with coefficients in $K(a)$ of degree > 0. The main technique in the proof is to take the *norm* of $f(x)$ with respect to $K(a)$ over K —which is a polynomial with coefficients in K —and to factor it. Loosely speaking, the norm $Nf(x)$ of $f(x)$ is obtained by replacing each a which occurs in $f(x)$ successively by $a', a'', \ldots, a^{(n-1)}$, where these are the other roots of ϕ, and setting $Nf(x)$ equal to the product of these n polynomials; since the coefficients of $Nf(x)$ are symmetric in the roots $a, a', \ldots, a^{(n-1)}$ of ϕ, they are in K. However, this cannot be used as a definition of $Nf(x)$ because it assumes that there is some field (or at least some ring) containing all roots $a, a', a'', \ldots,$ of ϕ, in which the multiplication can take place, and the existence of such a field is a large part of what is to be proved. Thus another approach must be taken to the definition of $Nf(x)$.

Let $K(a)$ be regarded as a vector space of dimension n over K and let multiplication by an element of $K(a)$ be regarded as an automorphism of the vector space. Then this automorphism has a determinant and this determinant can be defined to be the norm of the corresponding element of $K(a)$. This definition has the disadvantages that it does not immediately give a definition of the norm of a *polynomial* $f(x)$ with coefficients in $K(a)$ and that it uses the language of vector spaces, with which the reader may not be familiar. Both of these objections can be removed by making the definition more explicit:

An element of $K(a)$ can, by the definition of $K(a)$, be written in just one way in the form $A_0 + A_1 a + A_2 a^2 + \cdots + A_{n-1} a^{n-1}$ where the A's are in K. For any given element of $K(a)$, say μ, let M_μ be the $n \times n$ matrix $M_\mu = (A_{ij})$ of elements of K defined by $\mu \cdot a^i = \sum_{j=0}^{n-1} A_{ij} a^j$ ($i = 0, 1, \ldots, n - 1$) and let the norm of μ be *defined* by $N(\mu) = \det(M_\mu)$. This definition of the norm of an element of $K(a)$ extends immediately to give norms of polynomials with coefficients in $K(a)$, because if $F(x, y, \ldots)$ is a polynomial in any number of variables with coefficients in $K(a)$ then the elements A_{ij} of M_F can be defined in the same way as before; the A_{ij} are polynomials in the same variables as occur in F, with coefficients in K, and the determinant of M_F is a polynomial in these variables with coefficients in K which is defined to be $N(F)$. It follows from the definition of matrix products that if F and G are polynomials with coefficients in K then $M_{FG} = M_F M_G$, and therefore, from the fact that the determinant of a product is the product of the determinants, that $N(FG) = N(F)N(G)$. Finally, if F has all its coefficients in K then A_{ij} is 0 if $i \neq j$ and is F if $i = j$, so that $N(F) = F^n$ in this case.

With $Nf(x)$ so defined, let its factorization into irreducible factors (as a polynomial with coefficients in K it can be factored) be $Nf(x) = F_1 F_2 \ldots F_v$. For each $i = 1, 2, \ldots, v$, let the Euclidean algorithm be used to find the greatest common divisor of f and F_i, which is a polynomial with coefficients in $K(a)$. If this is a proper divisor of f then f will have been factored. However, this may not produce a factorization of f even when one exists; for example, if f is any polynomial with coefficients in K which is irreducible over K then $Nf = f^n$ and the greatest common divisors are all f and are never proper divisors of f even though f may not be irreducible over $K(a)$. (The polynomial ϕ is irreducible over K but not over $K(a)$.) For such a polynomial, one might try applying the same technique to $f(x + a)$ instead of $f(x)$. This polynomial has its coefficients in $K(a)$ but not in K, so the factors of its norm will be different from $f(x + a)$, and the greatest common divisors now may well include a proper factor of $f(x + a)$; if so, then a factorization $f(x + a) = g(x)h(x)$ will have been found, and this will give a factorization $f(x) = g(x - a)h(x - a)$ as required. If this fails, one might try $f(x + 2a)$ or $f(x - a)$ or $f(x + 3a)$, etc. One of Kronecker's basic techniques in the theory of algebraic functions was the *method of undetermined coefficients*. For him it was natural, therefore, to consider, instead of $f(x + na)$ for various values of n, the polynomial $f(x + ua)$ where u is a *new variable*. When this is done, the above method does give a factorization of f unless f is irreducible.

Specifically, let $f(x + ua)$ be regarded as a polynomial in the two variables x and u with coefficients in $K(a)$. Then $Nf(x + ua)$ is a polynomial in x and u with coefficients in K. As was seen in the preceding article, such a polynomial can be factored into irreducible factors (assuming that polynomials in one variable can be factored over K), say $Nf(x + ua) = G_1(x, u)G_2(x, u) \ldots G_\mu(x, u)$. Now both $f(x + ua)$ and $G_i(x, u)$ can be regarded as polynomials in the one variable x with coefficients in the field $K(a, u)$ obtained by adjoining the indeterminate u to $K(a)$ (a transcendental extension of $K(a)$). As such, they have a greatest common divisor with coefficients in this field which can be found using the Euclidean algorithm, say $d_i(x)$. Then $f(x + ua) = d_i(x)q_i(x)$ where d_i and q_i have coefficients in $K(a, u)$. This equation can be multiplied by the product of the denominators of the coefficients of d_i and q_i to find an equation of the form $H_i(u)f(x + ua) = D_i(x)Q_i(x)$ where H_i is a polynomial in u with coefficients in $K(a)$, and where D_i and Q_i have coefficients that are elements of $K(a, u)$ without denominators, that is, polynomials in u with coefficients in $K(a)$. This means that D_i and Q_i can be regarded as polynomials in two variables, say $D_i(x, u)$ and $Q_i(x, u)$, with coefficients in $K(a)$. With $u = 0$ this will give a factorization of $f(x)$ over $K(a)$ unless $H_i(0) = 0$. However, if $H_i(0) = 0$ then $H_i(u)$ has no constant term, so $D_i(x, u)Q_i(x, u)$, when regarded as a polynomial in u that has coefficients that are polynomials in x, has no constant term; in this case either D_i or Q_i has no constant term and a factor of u can be cancelled on both sides. Thus it can be assumed that $H_i(0) \neq 0$. Then $f(x) = H_i(0)^{-1}D_i(x, 0)Q_i(x, 0)$ is a factorization of $f(x)$ over $K(a)$. It is a *proper* factorization unless deg $D_i(x, 0)$ = deg d_i is either 0 or deg f. Thus, *in order to prove that the method described above is a factorization method for polynomials over $K(a)$ it will suffice to show that if $f(x)$ has the property that the $d_i(x)$ all have degree 0 or deg f then f is irreducible over $K(a)$.*

Assume, therefore, that deg d_i is always 0 or deg f. It will first be shown that, for at least one value of i, deg d_i = deg f. For each i there are polynomials A_i, B_i with coefficients in $K(a, u)$ such that $d_i(x) = A_i(x)f(x + ua) + B_i(x)G_i(x, u)$. If deg $d_i = 0$ in all cases, then the product of these expressions for d_i would give an element of $K(a, u)$ on the left and an expression of the form $Af + BG_1G_2 \ldots G_\mu$ on the right (the only term of the product without an f has the product of the G's). Since $G_1G_2 \ldots G_\mu = Nf$, this would show that the greatest common divisor of $f(x + ua)$ and $Nf(x + ua)$, when they are regarded as polynomials in x with coefficients in $K(a, u)$, was 1. To prove that this is impossible, it will suffice to prove that $f(x + ua)$ divides $Nf(x + ua)$, a property of the norm that would be immediate if it were defined as the product of the conjugates of f.

Lemma. *If $F(x, y, z, \ldots)$ is a polynomial in any number of variables with coefficients in $K(a)$, and if $N(F)$ is defined as above, then F divides $N(F)$ when both are regarded as polynomials with coefficients in $K(a)$.*

PROOF. Let X be a new variable and consider the norm of the polynomial $F - X$. This can be expanded as a polynomial in X, say $N(F - X) = A_0 X^n + A_1 X^{n-1} + \cdots + A_n$ where the A_i are polynomials in the same variables as F with coefficients in K. (It is clear from the definition that $N(F - X)$ has degree n in X and in fact that $A_0 = (-1)^n$.) With $X = 0$ this equation gives $N(F) = A_n$. With $X = F$ it gives $0 = A_0 F^n + A_1 F^{n-1} + \cdots + A_n$. (See Exercise 14.) Thus $N(F) = A_n = F(-A_0 F^{n-1} - A_1 F^{n-2} - \cdots - A_{n-1})$, which proves the lemma. $\square$

Thus there is at least one i for which $\deg d_i = \deg f$. It is to be shown that this implies that $f(x)$ is irreducible over $K(a)$. Suppose that $f(x) = g(x)h(x)$ where g and h have coefficients in $K(a)$. Then $f(x + ua) = g(x + ua)h(x + ua)$ and $Nf(x + ua) = Ng(x + ua)Nh(x + ua)$. Since $G_i(x, u)$ is an irreducible factor of $Nf(x + ua)$, and since factorization of polynomials in two variables with coefficients in a field is unique (Exercise 3), G_i must divide $Ng(x + ua)$ or $Nh(x + ua)$. Suppose it divides $Ng(x + ua)$. Since d_i divides both $f(x + ua)$ and G_i, and since $\deg f = \deg d_i$, $f(x + ua)$ divides G_i and therefore divides $Ng(x + ua)$ when these are regarded as polynomials in x with coefficients in $K(a, u)$. It will suffice to show that $f(x + ua)$ divides $Ng(x + ua)$ only when $\deg g = \deg f$ because this will show that the only factorizations $f = gh$ of f are the trivial ones.

Let $j = \deg g$ and $k = \deg f$. Since $f(x + ua)$ divides $Ng(x + ua)$ as a polynomial in x with coefficients in $K(a, u)$, there is an equation of the form

$$H(u) \cdot Ng(x + ua) = f(x + ua)Q(x, u),$$

where H clears out the denominators and where all terms are polynomials with coefficients in $K(a)$. Consider the terms of the highest combined degree in this equation. In $f(x + ua)$ the terms of highest degree are an element of K times $(x + ua)^k$. Thus $(x + ua)^k$ divides the terms of highest degree on the left. In $H(u)$ the term of highest degree is a nonzero element of $K(a)$ times a power of u, say u^r. The terms of highest degree in the matrix $M_{g(x + ua)}$ are of degree j and come from the leading term of $g(x + ua) = c(x + ua)^j + \cdots$. Therefore $Ng(x + ua) = \det M_{g(x + ua)} = N(c(x + ua)^j) + $ terms of degree less than nj. Since the terms $N(c(x + ua)^j) = N(c)N(x + ua)^j$ of degree nj are nonzero ($1 = N(1) = N(c)N(c^{-1})$ and $N(x + ua) = x^n + \cdots$) these are the terms of highest degree in $Ng(x + ua)$. Therefore, as a polynomial in x and u with coefficients in $K(a)$, $(x + ua)^k$ divides $u^r N(x + ua)^j$. It is to be shown that this implies that $j \geq k$ and therefore that $j = k$.

Set $u = -1$. Then $(x - a)^k$ divides $N(x - a)^j$. By the lemma above, $x - a$ divides $N(x - a)$, say $N(x - a) = (x - a)q(x)$. Then $(x - a)^k$ divides $(x - a)^j q(x)^j$, which is to say that $(x - a)^{k-j}$ divides $q(x)^j$. Unless $k = j$, it follows that $x - a$ divides $q(x)$, that is, that $(x - a)^2$ divides $N(x - a)$. It will suffice, therefore, to prove that *if a is the root of an irreducible polynomial* $\phi(a) = 0$ *over K then* $(x - a)^2$ *does not divide* $N(x - a)$ when both are regarded as polynomials with coefficients in $K(a)$.

This can be proved as follows. Taking the norm of the equation $\phi(x) = (x - a)q(x)$ shows that $\phi(x)^n = N(x - a) \cdot Nq$. By the unique factorization of polynomials with coefficients in K and the fact that ϕ is irreducible, it follows that $N(x - a)$ is a nonzero element of K times a power of $\phi(x)$. Since both are of degree n, $N(x - a)$ is a nonzero element of K times $\phi(x)$, and it suffices to show that $(x - a)^2$ does not divide $\phi(x)$. If $(x - a)^2$ did divide $\phi(x)$ then differentiation* of the equation $\phi(x) = (x - a)^2 q_2(x)$ would show that $\phi'(x)$ had the factor $x - a$ in common with $\phi(x)$. Since $\phi(x)$ is irreducible, it would follow (Galois' Lemma I) that $\phi(x)$ divided $\phi'(x)$, which is impossible because $\deg \phi = n \geq 2$ and $\deg \phi' = n - 1 > 0$.   □

§61 This completes the proof that polynomials with coefficients in K can be factored whenever K is a field obtained from the rational field $\mathbb{Q}$ by a finite number of algebraic or transcendental adjunctions. Three remarks are in order in connection with this proof. The first is that, although most of the ideas are to be found in Kronecker's Grundzüge ([K4], §4), the one really hard part—namely, the proof of the preceding article that if the greatest common divisors of $f(x + ua)$ and the irreducible factors of $Nf(x + ua)$ do not give a factor of $f(x)$ over $K(a)$ then $f(x)$ is irreducible over $K(a)$—is not to be found in Kronecker. He merely says that this is "easy to see". The basic idea of the proof given above is taken from van der Waerden [W2]. By the way, Kronecker motivates the introduction of $f(x + ua)$ in much the same way that this was done above, saying that it is necessary to insure that the coefficients involve a.

Second, if F_p denotes the field of integers modulo a prime p, then there is a factorization method for polynomials with coefficients in F_p because there are only a finite number of factors to try. All of the above arguments apply to give a factorization method for polynomials over fields obtained from F_p by a finite number of transcendental or algebraic adjunctions *except* the very last step, where it is stated that ϕ' cannot be divisible by ϕ. In fact, if ϕ is a polynomial of the form $\phi(x) = \psi(x^p)$, then $\phi'(x) = \psi'(x^p) \cdot px^{p-1}$ and, since $p = 0$ in F_p or in any extension of F_p, $\phi'(x) = 0$ and ϕ' is divisible by ϕ. The proof applies under the *additional assumption* that $\phi(x)$ does not

---

* The *derivative* of a polynomial $g(x)$ can be defined, without using limits, as the polynomial $g'(x)$ obtained by expanding $g(x + h)$ in the form $g(x + h) = A(x) + B(x)h + C(x)h^2 + \cdots$ and setting $B(x) = g'(x)$. The basic differentiation formulas $(cg)' = cg'$, $(g + h)' = g' + h'$, $(gh)' = g'h + gh'$ are all easily established. The elementary facts concerning formal differentiation of polynomials with coefficients in a given field are covered in Exercise 2 of the Fourth Set.

divide $\phi'(x)$. A simple algebraic extension is called *separable* if ϕ does not divide ϕ', where ϕ is the irreducible polynomial of which a root is being adjoined. (The separability condition can also, by Exercise 10, be stated in any of the following three forms: (1) $\phi' \neq 0$. (2) ϕ is not a polynomial in x^p. (3) There is a splitting field for ϕ in which it has deg ϕ distinct roots.) The above arguments show that if K is any field obtained from F_p by a finite number of transcendental or *separable* algebraic extensions then polynomials with coefficients in K can be factored. This suffices, by the argument of §51, to construct a *splitting field* for any polynomial with coefficients in such a field K, provided the polynomial has distinct roots (i.e. is relatively prime to its derivative). As will be seen in Exercises 10–15 of the Eighth Set, Galois theory applies, with minor modifications, to the solution of equations with coefficients in such a field K.

Finally, it must be remarked that the computations required to factor polynomials—and therefore to construct the splitting field and the Galois group—are too long to be carried out except in very simple cases.* Kronecker at one point (*loc. cit.*) says that the computation can be done "theoretically." Galois puts it more pungently. "If now you give me an equation that you have chosen at your pleasure, and if you want to know if it is or is not solvable by radicals, I need do nothing more than to indicate to you the means of answering your question, without wanting to give myself or anyone else the task of doing it. In a word, the calculations are impractical." (Galois, [G1], p. 39.)

Seventh Exercise Set

1. Given the theorem of §54, prove that if $f_1 f_2 \ldots f_\mu = g_1 g_2 \ldots g_\nu$ where the f's and g's are all irreducible, then $\mu = \nu$ and the g's can be reordered, if necessary, so that f_i is a unit times g_i for all $i = 1, 2, \ldots, \mu$.

2. Deduce Corollary 1 of §57 from the theorem.

3. Prove:

Theorem. *Let K be a field over which polynomials can be factored. In the ring $K[a, x]$ of polynomials in two variables with coefficients in K, irreducible elements are prime.*

Corollary 1. *If $F(a, x) \in K[a, x]$ is irreducible and if it has degree > 0 in x, then it is irreducible when considered as a polynomial in x with coefficients in $K(a)$.*

Corollary 2. *A representation of an element of $K[a, x]$ as a product of irreducibles is unique up to the order of the factors and multiplication by units (nonzero elements of K).*

4. Show that a polynomial in x with coefficients in $K(a)$ is a unit (that is, a nonzero element of $K(a)$) times a polynomial in two variables. Show also that a factorization of a polynomial into irreducible factors implies such a factorization of any unit times the polynomial.

---

* Of course much larger cases can be handled with modern computing machines. Algorithms for factoring polynomials are being studied very actively today, and Galois' imaginary calculations are becoming more and more feasible.

5. In the notation of §59, give an algorithm for computing g when $\tilde{g}$ is given.

6. Factor $2x^4 + 8x^3 + 9x^2 + 2x - 3$.

7. Let $K = \mathbb{Q}(a)$ be the simple algebraic extension of $\mathbb{Q}$ obtained by adjoining a root a of the irreducible polynomial $x^3 + x + 2$. Find the norm of $2a + 1$. Show that this is the number obtained by forming the product of the three conjugates of $2a + 1$ in the splitting field and expressing this as a symmetric polynomial in the roots.

8. Prove the following generalization of Gauss's lemma:

If g and h are polynomials with rational coefficients and if gh has integer coefficients, then any coefficient of g times any coefficient of h is an integer.

9. Show that the norm of a polynomial with coefficients in $K(a)$ can also be defined as the product of its conjugates. That is, let f be a polynomial in any number of variables with coefficients in $K(a)$, where a is algebraic over a field K in which polynomials can be factored, let ϕ be the irreducible polynomial with coefficients in K of which a is a root, and let $L = K(a, b, c, \ldots)$ be a splitting field of ϕ in which $a, b, c, \ldots$ are the roots of ϕ. Let f_b, $f_c, \ldots$ be the polynomials obtained by changing all a's in the coefficients of f to b, $c, \ldots$, respectively, and let $g = f f_b f_c \cdots$. Then $g = Nf$. [First consider the case $L = K(a)$. Then $Ng = Nf \cdot Nf_b \cdot Nf_c \cdots$. On one hand $Ng = g^n$, and on the other hand $Nf \cdot Nf_b \cdot Nf_c \cdots = Nf^n$. Unique factorization of polynomials then implies that $g = CNf$ where $C \in K$ satisfies $C^n = 1$. Applying this to $X - f$ instead of f gives $X^n + \cdots \pm g = C(X^n + \cdots \pm Nf)$, from which $g = Nf$. Then if $L = K(t)$, where t is a Galois resolvent, $N_t f = \prod f_{Sa} = g^k$ where N_t denotes the norm of a polynomial with coefficients in $L = K(t)$. The Galois group of $\phi(x) = 0$ over $K(a)$ has k elements because this is the number of elements in the Galois group that leave a fixed. Therefore t is the root of an irreducible polynomial of degree k with coefficients in $K(a)$. Therefore $a^i t^j$ for $0 \le i < n, 0 \le j < k$ are a basis of $K(t)$ over K. When this basis is used to compute the norm of a polynomial with coefficients in $K(a)$ it shows that $N_t f = (N_a f)^k$.]

10. Show that the three alternative definitions of separability given in §61 are equivalent to the definition that was given. [Note that in any extension of F_p, the identity $(x + y)^p = x^p + y^p$ holds.]

11. A field of characteristic p (that is, a field which contains F_p as a subfield where p is a prime integer) is said to be *perfect* if every element of the field has a pth root. Prove:

(1) F_p is perfect.
(2) Every simple algebraic extension $K(a)$ of a perfect field K is perfect.
(3) Every algebraic extension of a perfect field is separable.
(4) A transcendental extension $K(a)$ of a perfect field K is *not* perfect.

12. Show that if K has characteristic p and contains an element b with no pth root, and if $K(a)$ is obtained from K by adjoining a root a of $x^p - b$ then the norm of a polynomial with coefficients in $K(a)$ is just its pth power.

13. The analog of the theorem of §34 for transcendental instead of algebraic extensions states: Let K be a field. Then there is a field $K(t)$ containing K and containing an element t such that:

(1) every element of $K(t)$ can be expressed rationally in terms of t and elements of K; and
(2) if $f(X)$ is a polynomial with coefficients in K and if $f(t) = 0$ in $K(t)$ then $f(X) = 0$.

Moreover, if $K(t)$ and $K(t')$ are any two such fields then there is an isomorphism $K(t) \rightarrow K(t')$ which carries elements of K to themselves and carries t to t'. Prove this theorem.

14. As in §60, let F be a polynomial (in any number of variables) with coefficients in an algebraic extension $K(a)$ of K. Let $N(X - F)$ be the norm of $X - F$ as it is defined in §60. This is a polynomial in the variables of F and in the additional variable X with coefficients in K. Show that substitution of F for X in this polynomial gives the zero polynomial.

Review

§62 It was shown in the preceding section that, if K is the field of rational numbers $\mathbb{Q}$, or any field obtained from $\mathbb{Q}$ by a finite number of adjunctions, either algebraic or transcendental, then there is a factorization method for polynomials with coefficients in K. By the argument of §51, it follows that if f is any polynomial with coefficients in K then there is a *splitting field* for f, that is, a field L containing K such that f is equal to a product of linear polynomials with coefficients in L. Thus $f(x) = k(x - a)(x - b)(x - c)\ldots$ where k is in K and $a, b, c, \ldots$ are in L.

Given a splitting field L for f, the field $K(a, b, c, \ldots)$ on which Galois theory is predicated exists as a subfield of L, and all the constructions of the theory can be carried out. It is easy to see that any two fields $K(a, b, c, \ldots)$ obtained in this way as subfields of splitting fields L of f over K are isomorphic (i.e. there is a one-to-one correspondence between the elements of the two fields in which sums correspond to sums, products to products, and elements of K to themselves). For example,* one can observe that the polynomial $F(X) = \prod (X - St)$ of degree $n!$ of which a Galois resolvent t (see §32) is a root has coefficients which are symmetric functions of the roots $a, b, c, \ldots$ of f with coefficients in K; therefore these coefficients of F are known elements of K independent of the choice of L. If t' is any one of the $n!$ roots of F in L, then t' is in the subfield $K(a, b, c, \ldots)$ of L and, by Lagrange's theorem, $K(t') = K(a, b, c, \ldots)$. Now t' is a root of an irreducible factor $G(X)$ of F over K and, by §34, the field $K(t')$ is isomorphic to the field obtained by adjoining a root of this irreducible polynomial to K. Since t' was arbitrary, this shows that $K(a, b, c, \ldots)$ is isomorphic to the field obtained by adjoining to K a root of any irreducible factor of F. Thus $K(a, b, c, \ldots)$ is independent of L, up to isomorphism. It is therefore legitimate to call this *the* splitting field of f over K.

The *Galois group* of $f(x) = 0$ over K can be defined as the group of all automorphisms† of the splitting field $K(a, b, c, \ldots)$ of f over K that leave all

---

* This method of proof assumes that the roots $a, b, c, \ldots$ are distinct or, what is the same, that the greatest common divisor of f and f'' has degree 0. As was observed in §31, if f does not have this property, then it is easy to construct another polynomial f^*, with the same roots as f, which does have this property.

† An automorphism is an isomorphism of a field with itself.

elements of K fixed. In terms closer to those Galois used, it is the group of all permutations of the roots $a, b, c, \ldots$ that can be extended to be automorphisms of the entire splitting field $K(a, b, c, \ldots)$ which leave all elements of K fixed.*

The Fundamental Theorem of Galois Theory

§63   The basic fact of Galois theory is that the elements of the Galois group, which leave elements of K fixed by definition, leave *only* the elements of K fixed, that is:

If an element of the splitting field $K(a, b, c, \ldots)$ is left fixed by all the automorphisms of the Galois group then it is in K.

This, in effect, is Galois' Proposition I (see §41).

It is only a short step from this proposition to what is often called the fundamental theorem of Galois theory:

Let K' be a field contained in the splitting field and containing K. Then the Galois group of $f(x) = 0$ over K' is a subgroup of the Galois group of $f(x) = 0$ over K. This assignment of subgroups to subfields is a one-to-one onto correspondence between such subfields K' and subgroups of the Galois group.

By Proposition I, the Galois group of $f(x) = 0$ over K' determines K' as the subfield of elements it leaves fixed. This shows that the correspondence is one-to-one, that is, that different subfields correspond to different subgroups. To show that it is onto, let G' be a subgroup of the Galois group, let K' be the subfield† of the splitting field left fixed by all elements of G', and let G'' be the subgroup corresponding to K'. Then $G' \subset G''$ because G'' contains *all* automorphisms of the splitting field which leave K' fixed. It was shown in §46 that if G' is a proper subgroup of the Galois group of $f(x) = 0$ over K' then there is an element *not* in K' that is left fixed by all automorphisms in G'. Under the assumptions that have been made, therefore, G' is not a proper subgroup, that is, $G' = G''$ and G' corresponds to the subfield K'. Therefore the correspondence is onto.   □

---

* Galois' description of the elements of the group as *substitutions of the roots*, as opposed to automorphisms of the splitting field, has the advantage that extension of the field K to a field K' reduces the Galois group to a subgroup. (See §43.) This is not the case when elements of the group are regarded as automorphisms of the splitting field because then the two groups contain automorphisms of *different fields*, $K(a, b, c, \cdots)$ and $K'(a, b, c. \cdots)$.

† If S is an element of the Galois group then it is clear that the elements of the splitting field that are left fixed by S form a subfield containing K. Thus the intersection of these subfields over any set of S's is a subfield containing K.

The usual statement of the "fundamental theorem of Galois theory" also includes statements about the *degrees* of the field extensions; these statements too follow easily from Galois' Proposition I.

The *degree* of a field extension $K' \supset K$, denoted $[K':K]$, is defined to be the dimension of K' as a vector space over K. This positive integer is defined as follows. A finite subset $s_1, s_2, \ldots, s_n$ of K' is said to *span* K' over K if every element of K' can be written as a linear combination $k_1 s_1 + k_2 s_2 + \cdots + k_n s_n$ with coefficients k_i in K. A subset $s_1, s_2, \ldots, s_n$ is said to be *linearly independent* over K if there is no nontrivial linear relation $k_1 s_1 + k_2 s_2 + \cdots + k_n s_n = 0$, with coefficients k_i in K, that is, if the only such relation is the one in which the k_i are all 0. A subset $s_1, s_2, \ldots, s_n$ is said to be a *basis* of K' over K if it *both* spans K' over K and is linearly independent over K. It is simple linear algebra (Exercise 1) to prove that if there is a finite subset of K' which spans K' over K then there is a basis of K' over K, any two bases of K' over K have the same number of elements, and the transition between two bases $s_1, s_2, \ldots, s_n$ and $s_1', s_2', \ldots, s_n'$ is effected by an invertible $n \times n$ matrix (k_{ij}) of coefficients from K in the formula $s_i' = \sum_{j=1}^{n} k_{ij} s_j$.

Given an equation $f(x) = 0$ with coefficients in K, let G be its Galois group and let L be its splitting field. It was seen above that intermediate fields K', $K \subset K' \subset L$, correspond one-to-one to subgroups G' of G. *In this correspondence* $[L:K']$ *is equal to the order of* G' *and* $[K':K]$ *is equal to the index of* G' *in* G. This is easily proved. First, since $L = K(t)$ and since $1, t, t^2, \ldots, t^{n-1}$ is a basis of $K(t)$ over K, where n is the degree of the irreducible polynomial with coefficients in K of which t is a root, $[L:K] = [K(t):K] = n$. But n is also the order of the Galois group. Thus $[L:K]$ is the order of the Galois group of $f(x) = 0$ over K. Thus $[L:K']$ is the order of G', because G' is the Galois group of $f(x) = 0$ over K'. Therefore, the index of G' in G, that is, the order of G divided by the order of G', is $[L:K]/[L:K']$. Thus the final statement to be proved is $[L:K] = [L:K'][K':K]$, which follows from the fact that if $s_1, s_2, \ldots, s_\mu$ is a basis of K' over K and $u_1, u_2, \ldots, u_m$ is a basis of L over K' then the $m\mu$ elements $s_i u_j$ are a basis of L over K (Exercise 2).   □

A final observation which will be useful in the applications is that if $f(x)$ has no multiple roots then $f(x)$ *is irreducible over* K *if and only if the Galois group of* $f(x) = 0$ *over* K *acts transitively on the roots*, that is, if and only if for any two roots a and b of $f(x) = 0$ in the splitting field there is an automorphism S in the Galois group with $S(a) = b$. For, if $f(x) = g(x)h(x)$ is a nontrivial factorization of f and if a and b are roots of g and h respectively then application of any element S of the Galois group to $g(a) = 0$ gives $g(Sa) = 0$ and shows that $Sa \neq b$; therefore the action of the Galois group is not transitive. Conversely, if the action of the Galois group is not transitive, say if a and b are roots of f and no element of the Galois group carries a to b, let $a_1, a_2, \ldots, a_\mu$ be the roots of f that are of the form Sa for some S; then $g(x) = (x - a_1)(x - a_2) \ldots (x - a_\mu)$ is a proper divisor of f (b is a root of

f but not of g) with coefficients in K (it is unchanged by the Galois group) so f is reducible over K.   □

Construction of pth Roots of Unity

§64   It was observed in §42 that the Galois group of $x^p - 1 = 0$ over $\mathbb{Q}$ (p a prime) is a subgroup of the group of cyclic permutations presented by

$$
\begin{matrix}
a_1 & a_2 & a_3 & \cdots & a_{p-1} \\
a_2 & a_3 & a_4 & \cdots & a_1 \\
\cdots & & & & \\
a_{p-1} & a_1 & a_2 & \cdots & a_{p-2},
\end{matrix}
\tag{1}
$$

where $a_1, a_2, \ldots, a_{p-1}$ are the roots of $(x^p - 1)/(x - 1) = x^{p-1} + x^{p-2} + \cdots + x + 1$ ordered by choosing a primitive root g mod p and setting $a_{i+1} = a_i^g$. Galois stated without proof that the Galois group in this case is equal to the *entire* $(p - 1)$-element cyclic group.

Since it is clear that no proper subgroup of (1) acts transitively on the a's, in order to prove Galois' statement it will suffice, by the last proposition of the preceding article, to prove that $f(x) = x^{p-1} + x^{p-2} + \cdots + x + 1$ *is irreducible* over $\mathbb{Q}$. This very important fact had been rigorously proved by Gauss in the *Disquisitiones Arithmeticae* (Art. 341), so of course Galois could take it as known. Although Gauss's proof is straightforward and not difficult, it can be simplified—or at least shortened—somewhat. The proof which follows is due to Kronecker [K3].*

Let L be a splitting field for $f(x) = x^{p-1} + \cdots + x + 1$ over $\mathbb{Q}$. If a is any root of f in L then $a \neq 1$ and $a^p - 1 = (a - 1)f(a) = 0$. For any positive integer j, it follows that $(a^j - 1)f(a^j) = a^{jp} - 1 = 1 - 1 = 0$ so that either $a^j = 1$ or $f(a^j) = 0$. But if $a^j = 1$ and $a^p = 1$ then p divides j because otherwise their greatest common divisor is $1 = Aj + Bp$, where A and B are integers, and $a = a^1 = (a^j)^A(a^p)^B = 1^A 1^B = 1$, contrary to assumption. Thus $a, a^2, a^3, \ldots, a^{p-1}$ are all roots of f. They are distinct because if $a^j = a^{j+k}$ then $a^j(a^k - 1) = 0$, and either $a = 0$ (which is impossible) or $a^k = 1$ (which implies that p divides k). Thus $f(x) = (x - a)(x - a^2) \ldots (x - a^{p-1})$ is a factorization of f over L, and in fact over $\mathbb{Q}(a)$.

If $g(x)$ is any polynomial with integer coefficients, then, by the fundamental theorem on symmetric functions, $g(a)g(a^2) \ldots g(a^{p-1})$ can be expressed as a polynomial in the coefficients of f with integer coefficients. Thus $g(a)g(a^2) \ldots g(a^{p-1})$ is an integer. Kronecker's proof is based on the fact that this integer is congruent to $g(1)^{p-1}$ mod p. To prove this fact, let $G(x) = g(x)g(x^2) \ldots g(x^{p-1})$. Then G is a polynomial with integer coefficients, say

---

* For a shorter but trickier proof see Exercise 8.

$G(x) = A_0 + A_1 x + A_2 x^2 + \cdots$. Now $G(1) + G(a) + G(a^2) + \cdots + G(a^{p-1})$ can be computed in two ways. On the one hand, it is equal to

$$A_0 \cdot p + A_1 \cdot 0 + A_2 \cdot 0 + \cdots + A_{p-1} \cdot 0 + A_p \cdot p + A_{p+1} \cdot 0 + \cdots,$$

because $1 + a^j + a^{2j} + \cdots + a^{(p-1)j} = f(a^j)$ is equal to p if $a^j = 1$ and is equal to 0 if $a^j \neq 1$. In particular, $G(1) + G(a) + \cdots + G(a^{p-1})$ is an integer divisible by p. On the other hand, it is $g(1)^{p-1} + (p-1)g(a)g(a^2) \ldots g(a^{p-1})$, because $G(1) = g(1)^{p-1}$ and $G(a^j) = G(a)$ for $j = 1, 2, \ldots, p-1$ because replacement of a by a^j merely permutes the factors of $G(a)$ (Exercise 3). Thus $g(1)^{p-1} + (p-1)g(a)g(a^2) \ldots g(a^{p-1}) \equiv 0 \bmod p$ and, as desired, $g(a)g(a^2) \ldots g(a^{p-1}) \equiv g(1)^{p-1} \bmod p$. If $f(x) = g(x)h(x)$ is a factorization of f, then, since f has leading coefficient 1, g and h can be multiplied by rational numbers to make their leading coefficients equal to 1 without changing their product $f = gh$. Then by Gauss's lemma (§57) g and h have *integer* coefficients. It will suffice to show that the only such factorizations are the trivial ones in which one factor is 1. Since $f(1) = p$, the factorization $f(1) = g(1)h(1)$ must be trivial, say $g(1) = \pm 1$. Then, by the lemma, $g(a)g(a^2) \ldots g(a^{p-1}) \equiv g(1)^{p-1} \equiv 1 \bmod p$. On the other hand, from $g(x)h(x) = f(x) = (x - a)(x - a^2) \ldots (x - a^{p-1})$, it follows from uniqueness of factorization of polynomials over $\mathbb{Q}(a)$ that $g(x)$ is an element of $\mathbb{Q}(a)$ times a product of factors $x - a^j$. If $\deg g > 0$ it would follow that $g(a^j) = 0$ for some j, from which $0 = g(a)g(a^2) \ldots g(a^{p-1}) \equiv g(1)^{p-1} \equiv 1 \bmod p$. Since this is impossible, $\deg g = 0$, as was to be shown. $\qquad\square$

§65 Since the Galois group of $x^p - 1 = 0$ over $\mathbb{Q}$ is the cyclic group of order $p - 1$, the problem of solving this equation by radicals amounts to the problem of finding a sequence of subgroups of this group to show that it is solvable. Such subgroups are easy to give explicitly. Let S_i denote the substitution which carries the 1st row of (1) to the $(i + 1)$st row, for $i = 0, 1, \ldots, p - 2$. Then S_i followed by S_j is equal to S_{i+j} whenever $i \geq 0$, $j \geq 0$, $i + j < p - 1$, and if S_i is defined for all integers i by setting $S_i = S_j$ whenever $i \equiv j \bmod p - 1$ then $S_i S_j = S_{i+j}$ for all integers i and j. In particular, these substitutions *commute*. If H is any subgroup, then its order divides $p - 1$, say $p - 1 = hq$ where h is the order of H. If S_i is any element of H then the powers of S_i form a subgroup of H and therefore the number of distinct powers of S_i divides h, that is, if j is the least integer such that $S_i^j = S_0$ then j divides h. Thus $S_i^h = S_0$. Since $S_i^h = S_{ih}$, it follows that ih is divisible by $p - 1 = qh$, which is to say that i is divisible by q. Thus S_i must be one of the h substitutions $S_q, S_{2q}, \ldots, S_{hq} = S_0$. This shows that *for each divisor h of $p - 1$ there is one and only one subgroup of order h, namely,*

$$\{S_q, S_{2q}, \ldots, S_{p-1} = S_0\}$$

where $q = (p - 1)/h$. Then a sequence of subgroups of the desired type $G \supset H_1 \supset H_2 \supset \cdots \supset H_v = \{S_0\}$ corresponds to a sequence of divisors $p - 1 > h_1 > h_2 > h_3 > \cdots > h_v = 1$ of $p - 1$ with the property that

h_i/h_{i+1} is prime. Such a sequence gives a prime factorization $p - 1 = p_1 p_2 \cdots p_v$ where $p_1 = (p - 1)/h_1, p_2 = h_1/h_2, \ldots, p_v = h_{v-1}/h_v$. Conversely, given such a prime factorization of $p - 1$, let H_k be the subgroup of G of order $p_{k+1}p_{k+2} \cdots p_v$. Then the sequence of subgroups

$$G = H_0 \supset H_1 \supset \cdots \supset H_v = \{S_0\}$$

shows that G is solvable because the commutativity of H_i implies that all its subgroups—and in particular H_{i+1}—are normal.

Thus the Galois group of $x^p - 1 = 0$ over $\mathbb{Q}$ is solvable, and, by basic Galois theory, the equation is solvable by radicals. There is one point that needs to be observed, however: the theorem of §47, which states that an equation with solvable Galois group is solvable by radicals, *assumes* Gauss's theorem that pth roots of unity can be obtained by adjoining radicals, and therefore cannot be used in a proof of Gauss's theorem. The proof of Gauss's theorem can be based instead on the proposition of §46 as follows.

For any positive integer n let $K^{(n)}$ be the splitting field of $f(x) = \prod_{p \leq n}(x^p - 1)$ over $\mathbb{Q}$, where p ranges over prime integers. Since $K^{(n)}$ contains pth roots of unity whenever $n \geq p$, in order to show that pth roots of unity can be expressed in terms of radicals it will suffice to show that $K^{(n)}$ can be obtained from $\mathbb{Q}$ by a finite number of adjunctions of radicals. Since $K^{(2)} = \mathbb{Q}$, this is true for $n = 2$. Suppose that $K^{(n-1)}$ can be so constructed. If n is not prime then $K^{(n-1)} = K^{(n)}$ has already been constructed. Suppose, therefore, that n is prime, say $n = p$. The Galois group of $x^p - 1 = 0$ over $K^{(p-1)}$ is a subgroup of its Galois group over $\mathbb{Q}$. All such subgroups have been described above, and it has been noted that they are all solvable. Since $K^{(p-1)}$ contains all the needed roots of unity, the proposition of §46 shows that the splitting field of $x^p - 1$ over $K^{(p-1)}$ can be constructed by the successive adjunction of a finite number of radicals to $K^{(p-1)}$, and in fact it gives a determination of specific adjunctions that will achieve this. Since the splitting field of $x^p - 1$ over $K^{(p-1)}$ is $K^{(p)}$, this gives the construction of $K^{(p)}$, as desired. (This is, in effect, the construction of Gauss that was described in §25.)

Solution by Radicals

§66 Let $f(x) = 0$ be an equation with coefficients in a field K. A *solution by radicals* of such an equation is a sequence of field extensions $K \subset K_1 \subset K_2 \subset \cdots \subset K_\mu$ in which, for each i, K_i is obtained by adjoining the p_ith root of an element of K_{i-1} to K_{i-1}, where p_i is a prime and K_{i-1} contains a primitive p_ith root of unity, and in which K_μ is a splitting field for f. By the proposition of §44, it is clear that if such a solution exists then the Galois group G of $f(x) = 0$ over K is solvable, that is, there is a sequence of subgroups $G = H_0 \supset H_1 \supset \cdots \supset H_\mu = \{\text{identity}\}$ such that H_i is a normal subgroup of H_{i-1} of index 1 or p_i where p_i is prime. Conversely, if there is such a sequence of subgroups of the Galois group, then the construction of the preceding article with K in place of $\mathbb{Q}$ can be used to adjoin all the pth roots of unity for primes p that divide the order of the Galois group. This adjunction to K may reduce the Galois group of $f(x) = 0$ to a subgroup, but, as was shown in §47, a subgroup of a solvable group is solvable. Thus the proposition of §46 can be applied, because the needed roots of unity are all present in K, to construct a splitting field by radicals. Thus:

Theorem. *An equation is solvable by radicals if and only if its Galois group is solvable.*

§67 The only example of a Galois group Galois himself gave, other than the Galois group of $x^p - 1 = 0$ over $\mathbb{Q}$, was the example of an equation with *literal* coefficients, which he stated has as its Galois group the full group of all $n!$ permutations of the n roots.

Theorem. *Let K be a field and let $K' = K(A_1, A_2, \ldots, A_n)$ be the field obtained by adjoining n indeterminates to K. Then the Galois group of the equation $x^n + A_1 x^{n-1} + A_2 x^{n-2} + \cdots + A_n = 0$ over K' is the full group of all $n!$ permutations of the roots.*

PROOF. Let $a_1, a_2, \ldots, a_n$ be another set of n indeterminates and let $L = K(a_1, a_2, \ldots, a_n)$. Then, as is easily proved by induction, elements of L can be written in the form p/q where p and q are polynomials in $a_1, a_2, \ldots, a_n$ with coefficients in K and $q \neq 0$. Two such representations p_1/q_1 and p_2/q_2 represent the same element of L if and only if $p_1 q_2 = p_2 q_1$. The $n!$ element group of permutations of $a_1, a_2, \ldots, a_n$ acts on L in the obvious way by permuting a's in a representation p/q because if $p_1 q_2 = p_2 q_1$ and if S is any permutation of $a_1, a_2, \ldots, a_n$ then clearly $Sp_1 \cdot Sq_2 = Sp_2 \cdot Sq_1$. Let $\sigma_1, \sigma_2, \ldots, \sigma_n \in L$ be the elementary symmetric polynomials in $a_1, a_2, \ldots, a_n$ and let K' be mapped to L by substituting $(-1)^j \sigma_j$ for A_j in elements of K'. This mapping is well defined because if P_1, Q_1, P_2, Q_2 are polynomials in the A's with $P_1 Q_2 = P_2 Q_1$ and if p_1, q_1, p_2, q_2 are the corresponding symmetric polynomials in the a's then of course $p_1 q_2 = p_2 q_1$. Moreover, the mapping is *one-to-one*, that is, if $p_1 q_2 = p_2 q_1$ then $P_1 Q_2 = P_2 Q_1$, because a symmetric polynomial can be represented in *only one* way as a polynomial in the σ's (Exercise 22 of the First Set). Also, this mapping $K' \to L$ carries elements of K to themselves. Thus the extension $K' \supset K$ can be regarded as being *contained* in L and A_j regarded as being *equal* to $(-1)^j \sigma_j$. Then the equation $x^n + A_1 x^{n-1} + \cdots + A_n = x^n - \sigma_1 x^{n-1} + \cdots \pm \sigma_n = (x - a_1)(x - a_2) \ldots (x - a_n)$ shows that L is a *splitting field* for this polynomial. Therefore $K'(a_1, a_2, \ldots, a_n) = L$ is *the* splitting field of this polynomial over K'. Since any permutation S of the a's acts on the splitting field leaving K' fixed, the theorem follows. □

(For another proof of this theorem see Exercise 7.)

Corollary. *The general nth degree equation is solvable by radicals for n ≤ 4.*

PROOF. Let G be the group of all twenty-four permutations of a, b, c, and d. Since the group of two permutations of a and b, and the group of six permutations of a, b, and c can be regarded as subgroups of G, and since a subgroup of a solvable group is solvable, it will suffice to show that G is solvable. This Galois does in the form of a tableau:

abcd	acdb	adbc
badc	cabd	dacb
cdab	dbac	bcad
dcba	bdca	cbda.

(See (3) of §40.) Here the set of all twelve arrangements presents a twelve-element normal subgroup of index 2 in S_4. The three sets of four arrangements present (in three ways) a four-element normal subgroup of index 3 in this twelve element group. The first two (or in fact any two) arrangements in the first set of four present a two-element normal subgroup of index 2 in the four element group. Finally, the identity is a normal subgroup of index 2 in this two element group. □

Corollary. *Any equation* of degree ≤ 4 is solvable by radicals.*

PROOF. One need only solve the general equation and then substitute the values of the coefficients in the general solution. □

Note that the ancient solution of the quadratic equation is a solution of this type.

Corollary. *In order to prove that the general nth degree equation is not solvable by radicals for n ≥ 5 it suffices to prove this for n = 5.*

PROOF. The group of permutations of five letters can be regarded as a subgroup of the group of permutations of n letters for $n ≥ 5$. Therefore the latter cannot be solvable if the former is not. □

Corollary. *The general nth degree equation is irreducible.*

PROOF. Its Galois group acts transitively on the roots. □

---

* It is assumed here, as elsewhere in the text, that the field K of known quantities is obtained from $\mathbb{Q}$ by the adjunction of a finite number of algebraic and/or transcendental elements.

Solvable Equations of Prime Degree

§68 Theorem (Galois). *If an irreducible equation has prime degree p and is solvable by radicals, then the roots of the equation can be ordered $a_1, a_2, \ldots, a_p$ in such a way that the substitutions S of the Galois group are all of the form $S(a_i) = a_{ri+s}$, where a_i is defined for all integers i by setting $a_i = a_j$ for $i \equiv j \bmod p$, where r and s are integers, and where $r \not\equiv 0 \bmod p$.*

Corollary. *The general 5th degree equation is not solvable by radicals. Therefore the general nth degree equation is not solvable by radicals for $n \geq 5$.*

DEDUCTION OF COROLLARY. It was shown in the previous article that the first statement implies the second. Since the general 5th degree equation is irreducible and of prime order, the theorem shows that if it were solvable by radicals then the substitutions in its Galois group would all have the form $S(a_i) = a_{ri+s}$. Since $r \not\equiv 0 \bmod 5$, r can have four values mod 5 and s can have five values mod 5, so there are only twenty different substitutions of this type. Therefore the Galois group would contain at most twenty substitutions, rather than the $120 = 5!$ that it does contain.   □

Corollary. *An irreducible equation of prime degree is solvable by radicals if and only if it has the property that all its roots can be expressed rationally in terms of any two of them.*

This rather strange corollary is the principal result of Galois' memoir. No doubt Galois was attempting to state a theorem which went beyond Abel's theorem on the unsolvability of the quintic and which showed the power of his techniques but did not refer to the Galois group in its statement. Unfortunately, it did not catch the fancy of his readers at the French Academy of Sciences. Its deduction from the theorem is left as an exercise (Exercise 4).

PROOF OF THE THEOREM. The first step in the proof is Galois' Proposition VI, which states that if $K \subset K_1 \subset K_2 \subset \cdots \subset K_\mu$ is a solution of $f(x) = 0$ by radicals, and if $f(x)$ is irreducible of prime degree p, then $f(x)$ is irreducible over all the intermediate fields $K, K_1, K_2, \ldots, K_{\mu-1}$ until it splits into linear factors in K_μ. More precisely, *in a field extension of the type which occurs in a solution by radicals*—say the adjunction of a qth root to a field which contains qth roots of unity ($q = $ prime)—*an irreducible polynomial either remains irreducible or splits into q factors of equal degree.* The proof of this is an easy adaptation of the arguments of §44 (Exercise 5). (Alternatively, this statement follows immediately from Dedekind's "reciprocity theorem" of Exercises 10 and 11 of the Sixth Set.) Thus when $f(x)$ is irreducible of prime degree p, it remains irreducible unless it splits into factors of degree 1, and the splitting into factors of degree 1 can occur only with the adjunction of a

pth root. This also shows that the next-to-last group in the sequence of sub-groups corresponding to the solution $K \subset K_1 \subset \cdots \subset K_\mu$, which is the group of automorphisms of the splitting field K_μ over $K_{\mu-1}$, is cyclic of order p. Thus the roots can be numbered $a_1, a_2, \ldots, a_p$ in such a way that

$$
\begin{array}{cccc}
a_1 & a_2 & \cdots & a_p \\
a_2 & a_3 & \cdots & a_1 \\
\multicolumn{4}{c}{\cdots\cdots\cdots\cdots\cdots} \\
a_p & a_1 & \cdots & a_{p-1}
\end{array}
$$

is a presentation of this group.

Let S be a substitution of the roots which occurs in the *preceding* subgroup, and let S be represented as a permutation of the integers mod p by setting $S(a_j) = a_{S(j)}$. Because the group presented above is a normal subgroup, the application of S to this presentation gives another presentation of the same group

$$
\begin{array}{cccc}
a_{S(1)} & a_{S(2)} & \cdots & a_{S(p)} \\
a_{S(2)} & a_{S(3)} & \cdots & a_{S(1)} \\
\multicolumn{4}{c}{\cdots\cdots\cdots\cdots\cdots} \\
a_{S(p)} & a_{S(1)} & \cdots & a_{S(p-1)}.
\end{array}
$$

The fact that the substitution which changes the 1st row to the 2nd row must also change the 1st row of the first presentation to another row of that presentation means that there is an $r \not\equiv 0 \bmod p$ such that $a_{S(2)} = a_{S(1)+r}$, $a_{S(3)} = a_{S(2)+r}, \ldots$, where the subscripts are interpreted mod p. Thus

$$S(2) \equiv S(1) + r,$$
$$S(3) \equiv S(2) + r \equiv S(1) + 2r,$$
$$S(4) \equiv S(3) + r \equiv S(1) + 3r,$$
$$\ldots,$$
$$S(j) \equiv S(1) + (j-1)r \equiv rj + s,$$
$$\ldots,$$

where $s = S(1) - r$.

Thus the preceding group contains only substitutions of the prescribed type. Now consider the group which precedes this one. For the same reason as before, if T is any substitution of this group, then the substitution which carries the arrangement $a_{T(1)}a_{T(2)} \cdots a_{T(p)}$ to $a_{T(2)}a_{T(3)} \cdots a_{T(1)}$ must be a substitution of the normal subgroup and therefore there must be integers r and s such that

$$T(2) \equiv rT(1) + s,$$
$$T(3) \equiv rT(2) + s \equiv r^2 T(1) + rs + s,$$
$$T(4) \equiv rT(3) + s \equiv r^3 T(1) + r^2 s + rs + s,$$
$$\ldots,$$
$$T(j) \equiv r^{j-1} T(1) + (r^{j-2} + r^{j-3} + \cdots + r + 1)s,$$
$$\ldots.$$

Since $T(1) \equiv T(p + 1) \equiv r^p T(1) + (r^{p-1} + r^{p-2} + \cdots + r + 1)s$, and since $r^p \equiv r$ by Fermat's theorem, $(1 - r)T(1) \equiv (r^{p-1} + r^{p-2} + \cdots + r + 1)s$. Multiply by $1 - r$ to find $(1 - r)^2 T(1) \equiv (1 - r^p)s \equiv (1 - r)s$. If $r \not\equiv 1$ then $T(1) \equiv (1 - r)^{-1}s$. But then in the same way, $T(2) \equiv T(p + 2) \equiv r^p T(2) + (r^{p-1} + \cdots + r + 1)s$, which leads to $T(2) \equiv (1 - r)^{-1}s \equiv T(1)$, which is impossible. Therefore $r \equiv 1$ and $T(2) \equiv T(1) + s$, $T(3) \equiv T(2) + s$, $T(4) \equiv T(3) + s$, etc. As was seen above, this implies that T is a substitution of this same linear form $T(j) = r'j + s'$.

In the same way, the group preceding this one can contain only linear substitutions, and the same applies to *all* preceding groups, including the original one, which is the Galois group of $f(x)$ over K.   □

The Galois Group of $x^n - 1 = 0$

§69 The cyclotomic equation $x^n - 1 = 0$ is one of the simplest and most important algebraic equations. Therefore its Galois group is of particular interest. In the case where n is a prime p, it was seen above that, as Galois stated, this group is a cyclic group of order $p - 1$. For general n, the answer is easy to guess but not at all easy to prove:

By regarding roots of $x^n - 1 = 0$ as complex numbers, one sees that there is always a *primitive* root $a = \cos(2\pi/n) + i\sin(2\pi/n)$, that is, a root a with the property that $a^i \neq 1$ for $i = 1, 2, \ldots, n - 1$. Then $x^n - 1 = (x - a)(x - a^2) \cdots (x - a^n)$ because $a, a^2, \ldots, a^n = 1$ are n roots of $x^n - 1 = 0$ and they are distinct because if i and j are positive integers with $a^{i+j} = a^i$ then $a^j = 1$, which implies $j \geq n$. Thus the elements S of the Galois group can be regarded as permutations of the roots $a, a^2, a^3, \ldots, a^n = 1$. Since $S(a^i) = S(a)^i$, S is known once $S(a)$ is known. Suppose $S(a) = a^j$ $(1 \leq j \leq n)$. Then j must be relatively prime to n because if $d > 1$ divided both j and n, say $j = dJ$, $n = dN$, it would follow that $S(a^N) = S(a)^N = a^{jN} = a^{Jn} = 1 = S(1)$, contrary to the fact that $a^N \neq 1$ $(N < n)$ and S is one-to-one. With this limitation, that j be relatively prime to n, the permutations $a \mapsto a^j$ form a group as follows.

Let a be a symbol and let the 1st row of a table consist of the symbols a^j where j ranges over all integers $1 \leq j \leq n$ that are relatively prime to n. Let the table consist of one row for each entry in the 1st row, and let the row corresponding to a^j be obtained by replacing a with a^j in the 1st row and reducing all exponents mod n so they lie between 0 and n. (See §42 for this table in the case $n = 8$.) This table is a presentation of a group which is commonly denoted by $(\mathbb{Z}/n\mathbb{Z})^*$, the multiplicative group of invertible classes of integers mod n. It was seen above that *the Galois group of $x^n - 1 = 0$ over $\mathbb{Q}$ is in a natural way a subgroup of $(\mathbb{Z}/n\mathbb{Z})^*$.* In the case where n is prime, the Galois group was seen to be[†] all of $(\mathbb{Z}/n\mathbb{Z})^*$. In view of this, and in view

---

[†] The group is the same despite the fact that the presentation of $(\mathbb{Z}/n\mathbb{Z})^*$ that is described above is not the same as (1) of §64 because the rows and columns are ordered differently.

of the fact that there is no obvious way to pick out a subgroup of $(\mathbb{Z}/n\mathbb{Z})^*$, it is reasonable to guess that *in all cases the Galois group of $x^n - 1 = 0$ over $\mathbb{Q}$ is all of $(\mathbb{Z}/n\mathbb{Z})^*$*. This is the theorem to be proved.

§70 **Theorem.** *The splitting field of $x^n - 1$ over $\mathbb{Q}$ is of the form $\mathbb{Q}(a)$ where a is a primitive nth root of unity (that is, $a^n = 1$ but $a^i \neq 1$ for $i = 1, 2, \ldots, n - 1$). The primitive nth roots of unity in $\mathbb{Q}(a)$ are the elements a^j where $1 \leq j \leq n$ and j is relatively prime to n. Finally, the elements of the Galois group, which must carry a to other primitive nth roots of unity, carry a to all other primitive nth roots of unity a^j. Thus the Galois group is $(\mathbb{Z}/n\mathbb{Z})^*$.*

PROOF. It must first be shown that the splitting field of $x^n - 1$ over $\mathbb{Q}$ contains a primitive nth root of unity. This can be done by induction on the number of distinct prime factors of n as follows.

First suppose n has only one prime factor, that is, $n = p^k$ where p is prime and $k > 0$. Then $x^n - 1$ has the factorization $(x^{n/p} - 1)g(x)$ where $g(x) = x^{n-(n/p)} + x^{n-2(n/p)} + \cdots + x^{n/p} + 1$. The splitting field of $x^n - 1$ contains a root b of $g(x)$. Let i be the least positive integer such that $b^i = 1$. Then $i \leq n$ and i divides n because $n = qi + r$ where $0 \leq r < i$; since $1 = b^n$ (because b is a root of g and therefore of $x^n - 1) = b^{qi+r} = (b^i)^q b^r = b^r$ (because $b^i = 1$ by assumption) the definition of i is contradicted unless $r = 0$. Since $n = p^k$, it follows that i is a power of p. If i were not $n = p^k$ then it would divide $n/p = p^{k-1}$, which would imply $b^{n/p} = 1$ and substitution of b in $g(x)$ would give $1 + 1 + \cdots + 1 = p$, contrary to the definition of b. Therefore $i = n$, that is, b is a primitive nth root of unity.

Now suppose it has been shown that the splitting field of $x^n - 1$ over $\mathbb{Q}$ contains a primitive nth root of unity whenever n has fewer distinct prime factors. Specifically, let $n = mp^k$ where m is not divisible by p and suppose the splitting field of $x^m - 1$ over $\mathbb{Q}$ contains a primitive mth root of unity, call it c. Since the splitting field of $x^n - 1$ over $\mathbb{Q}$ contains splitting fields of both $x^m - 1$ and $x^{p^k} - 1$ over $\mathbb{Q}$, this field contains both a primitive p^kth root of unity b and a primitive mth root of unity c. Let $a = bc$. Then a is a primitive nth root of unity. This can be proved as follows. Clearly a is an nth root of unity because both b and c are. It is to be shown that if $a^i = 1$ then i is divisible by n. The Euclidean algorithm can be used to write $1 = Am + Bp^k$ because 1 is the greatest common divisor of m and p^k. If $a^i = 1$ is raised to the mth power one finds $1 = b^{mi}c^{mi} = b^{mi}$. This implies that p^k divides mi and therefore that $n = p^k m$ divides $mi = Ammi + Bp^k mi$ (n divides both terms on the right). Similarly, raising $a^i = 1$ to the p^kth power shows that $p^k i$ is divisible by n. Therefore $i = Ami + Bp^k i$ is divisible by n, as was to be shown.

Therefore the splitting field of $x^n - 1$ over $\mathbb{Q}$ is of the form $\mathbb{Q}(a)$, where a is a primitive nth root of unity. ($a, a^2, \ldots, a^n = 1$ are distinct roots of $x^n - 1 = 0$.) Every primitive nth root of unity is an nth root of unity and therefore has the form a^j for some j. As was noted in §69, if a^j is a primitive nth root of unity then j is relatively prime to n. Conversely, if j is relatively prime to n then $1 = An + Bj$ for some integers A and B and therefore $a = (a^n)^A(a^j)^B = (a^j)^B$; thus every power of a can be expressed as a power of a^j (negative powers of a are equal to positive powers of a^{n-1}) and a^j is a primitive nth root of unity. Thus the primitive roots of unity are the powers a^j of a, $1 \leq j \leq n$, in which j is relatively prime to n. The last and main statement of the theorem is that *for every such j there is an element of the Galois group of $x^n - 1$ over $\mathbb{Q}$ which carries a to a^j.*

Let $x^n - 1$ be factored into irreducible polynomials over $\mathbb{Q}$. Since $x^n - 1$ has leading coefficient 1, one can assume without loss of generality that all factors have leading coefficient 1. Then, by Gauss's lemma (§57), all the irreducible factors of $x^n - 1$ have coefficients in $\mathbb{Z}$. Let a be a primitive nth root of unity in the splitting field of $x^n - 1$ over $\mathbb{Q}$ and let $f(x)$ be the irreducible factor of $x^n - 1$ of which a is a root. The Galois group of $x^n - 1$ over $\mathbb{Q}$ acts transitively on the roots of $f(x)$ since otherwise the method at the end of §63 would give a proper divisor of f. Thus the main statement of the theorem amounts to saying that a^j is a root of $f(x)$ for every j relatively prime to n.

Proposition. *Let a be a primitive nth root of unity (in some algebraic extension of $\mathbb{Q}$) and let $f(x)$ be an irreducible polynomial with coefficients in $\mathbb{Z}$ and leading coefficient 1 of which a is a root. Let q be a prime integer which does not divide n. Then a^q is a root of $f(x)$.*

This appears at first to be weaker than the preceding statement because j was merely assumed to be *relatively* prime to n, whereas q is assumed to be prime. However, any j relatively prime to n is of the form $j = p_1 p_2 \cdots p_\nu$ where the p's are prime and do not divide n. By the proposition, a^{p_1} is a root of $f(x)$. Since it is also a primitive nth root of unity, the proposition implies then that $(a^{p_1})^{p_2}$ is a root of $f(x)$. Repetition of this argument ν times shows that $a^{p_1 p_2 \cdots p_\nu} = a^j$ is a root of $f(x)$. Therefore the theorem will follow once the proposition is proved.

PROOF OF PROPOSITION (Dedekind, 1857). Let $f_1(x)$ be the irreducible factor of $x^n - 1$ with leading coefficient 1 of which a^q is a root. If b is any other root of $f(x)$ then, as was noted above, there is an automorphism in the Galois group of $x^n - 1$ over $\mathbb{Q}$ which carries a to b. Since such an automorphism carries roots of f_1 to roots of f_1 and carries a^q to b^q, b^q is a root of f_1. Therefore the qth power of any root of f is a root of f_1. Conversely, since there are integers A and B with $Aq + Bn = 1$, $a = a^{Aq + Bn} = a^{Aq}$, the same argument applied to roots of f_1 shows that the Ath power of any root of f_1 is a root of f. Therefore f and f_1 have the same number of roots, and if

$$f(x) = (x - a)(x - b) \ldots (x - d)$$

is the splitting of f over $\mathbb{Q}(a)$ then

$$f_1(x) = (x - a^q)(x - b^q) \ldots (x - d^q)$$

is the splitting of f_1. Thus the coefficients of f_1 are the elementary symmetric functions in $a^q, b^q, \ldots, d^q$.

The main trick in Dedekind's proof is to use Fermat's theorem (Exercise 9) in the form of the statement that if $F(X, Y, \ldots, Z)$ is a polynomial in any number of variables with coefficients in $\mathbb{Z}$ then

$$F(X, Y, \ldots, Z)^q = F(X^q, Y^q, \ldots, Z^q) + qQ(X, Y, \ldots, Z)$$

where Q is a polynomial with integer coefficients. When this formula is applied to an elementary symmetric polynomial in the roots $a, b, c, \ldots$ of f, Q is a symmetric polynomial in $a, b, c, \ldots$ and therefore is an integer by the fundamental theorem on symmetric functions. Therefore each coefficient of $f_1(x)$ differs by an integral multiple of q from the corresponding coefficient of f. In short,

$$f_1(x) = f(x) + q\,h(x),$$

where $h(x)$ is a polynomial with integer coefficients.

Now if a^q is not a root of f then f and f_1 are distinct factors in the irreducible factorization of $x^n - 1$. If $f_1(x)$ is replaced by $f(x) + qh(x)$ in this factorization the result is an equation of the form

$$x^n - 1 = f(x)^2\phi(x) + q\psi(x),$$

where $\phi(x)$ and $\psi(x)$ are polynomials with integer coefficients. Differentiation of this equation gives one of the form

$$nx^{n-1} = f(x)\Phi(x) + q\Psi(x), \tag{1}$$

where $\Phi(x)$ and $\Psi(x)$ have integer coefficients. Very briefly, an equation of this form is impossible because it says that $f(x)$ divides a nonzero constant times a power of the irreducible polynomial x mod q; by the unique factorization of polynomials mod q, this would imply $f(x)$ was a nonzero constant times a power of x mod q, and this contradicts the fact that $f(x)$ divides $x^n - 1$.

To spell out the argument that (1) is impossible a little more fully, note first that one can assume without loss of generality that all coefficients A_i of Φ lie in the range $0 \leq A_i < q$ (because $\Phi(x)$ can be changed to $\Phi(x) - q\Delta(x)$ and $\Psi(x)$ can be changed to $\Psi(x) + f(x)\Delta(x)$ to give another equation of the same form in which the A_i are altered by arbitrary multiples of q). The leading term of $f(x)$ is 1, so the leading term of $f(x)\Phi(x)$ is $A_\mu x^{\mu + \nu}$ where μ is the degree of Φ and ν the degree of f. If $\mu + \nu \geq n$, (1) would give $0 = A_\mu$ plus a multiple of q, which is impossible. If $\mu + \nu \leq n - 2$ it would give $n = 0$ plus a multiple of q, which is also impossible. Therefore $\mu + \nu = n - 1$. On the other hand, since $f(x)$ divides $x^n - 1$, its constant term must divide -1 (set $x = 0$) that is, its constant term is ± 1. If τ is the least index such that $A_\tau \neq 0$, and if $\tau < n - 1$ then (1) gives $0 = \pm A_\tau$ plus a multiple of q, which is impossible. Thus $\tau \geq n - 1$, which contradicts $\tau \leq \mu = n - 1 - \nu < n - 1$.

Therefore a^q must be a root of $f(x)$ and the proposition is proved.   □

§71   The theorem of the preceding article can be used to plug the gap in Gauss's proof that $x^p = 1$ can be solved by radicals, because it implies Lemma 2 of §24, that is:

Corollary. *Let p be a prime and let $n = p(p - 1)$. The splitting field of $x^n - 1$ over $\mathbb{Q}$ contains both a primitive pth root of unity, say α, and a primitive $(p - 1)$st root of unity, say β. If $P_1(\beta)\alpha + P_2(\beta)\alpha^2 + \cdots + P_{p-1}(\beta)\alpha^{p-1} = 0$ where the P's are polynomials in β with rational coefficients then $P_1(\beta) = P_2(\beta) = \ldots = P_{p-1}(\beta) = 0$.*

DEDUCTION OF COROLLARY. Let $a = \alpha\beta$. Since p and $p - 1$ are relatively prime, it was shown above that a is a primitive nth root of unity ($n = p(p - 1)$). The idea of the proof is to consider the Galois group of $x^n - 1 = 0$ over the field $\mathbb{Q}(\beta)$. The elements of this group are those elements of the whole Galois group of $x^n - 1 = 0$ over $\mathbb{Q}$ which leave β fixed. Such an element carries a to a^j and carries β to itself. Since $a \to a^j$ and β is a power of a, $\beta \mapsto \beta^j$. Since β is a primitive $(p - 1)$st root of unity and $\beta^j = \beta$, the elements of the Galois group of $x^n - 1$ over $\mathbb{Q}(\beta)$ are those elements $a \mapsto a^j$ of the whole group for which j is 1 more than a multiple of $p - 1$ and, of course, $1 \leq j < n$ and j is relatively prime to $n = p(p - 1)$. The values of j are thus 1, $1 + (p - 1)$, $1 + 2(p - 1)$, $1 + 3(p - 1), \ldots, 1 + (p - 1)(p - 1)$—that is, 1, p, $2p - 1$, $3p - 2, \ldots, p^2 - 2p + 2$—except for the second value p, which is not relatively prime to n. Therefore *the Galois group of $x^n - 1 = 0$ over $\mathbb{Q}(\beta)$ has exactly $p - 1$ elements.* This implies (§63) that $[\mathbb{Q}(a) : \mathbb{Q}(\beta)] = p - 1$. But a relation of the form

$$\sum_1^{p-1} P_i(\beta)\alpha^i = 0$$

implies $\sum_1^{p-1} P_i(\beta)\beta^{1-i}a^{i-1} = 0$. If at least one of the P's were nonzero and if k were the largest integer for which $P_k(\beta) \neq 0$, then one could move the term with $i = k$ to the other side of the equation and divide by its coefficient to express a^{k-1} as a combination of lower powers of a and elements of the field $\mathbb{Q}(\beta)$. This would imply that every element of $\mathbb{Q}(a)$ could be expressed as a linear combination of $1, a, a^2, \ldots, a^{k-2}$ with coefficients in $\mathbb{Q}(\beta)$ and therefore would imply $[\mathbb{Q}(a) : \mathbb{Q}(\beta)] \leq k - 1 < k \leq p - 1$, contrary to what was found above. Therefore the P's must all be 0.   □

Eighth Exercise Set

1. A vector space V over a field K is a set whose elements can be added to each other and multiplied by elements of K in such ways that the natural axioms apply: $v_1 + v_2 = v_2 + v_1, (v_1 + v_2) + v_3 = v_1 + (v_2 + v_3), k(v_1 + v_2) = kv_1 + kv_2, (k_1 + k_2)v = k_1v +$

k_2v, and, lastly, any equation of the form $v = k_1v_1 + k_2v_2 + \cdots + k_mv_m$ in which $v, v_2, v_3, \ldots, v_m$ are in V; $k_1, k_2, \ldots, k_m$ are in K, and $k_1 \neq 0$, has a unique solution v_1 in V. Show that if a finite subset of V spans V then V has a finite basis, any two bases have the same number of elements, and the transition from one basis to another is given by an invertible square matrix of elements of K. (See §63 for definitions.)

2. If $K \subset K' \subset L$ are three fields, if $s_1, s_2, \ldots, s_n$ is a basis of K' over K, and if $u_1, u_2, \ldots, u_m$ is a basis of L over K', show that the mn elements s_iu_j are a basis of L over K.

3. In the notation of §64, show that $G(a^j) = G(a)$ when j is not divisible by p.

4. Deduce the second corollary to the theorem of §68. [Galois' indication of a proof (Appendix 1) seems difficult to complete. Good use can be made of "Cauchy's Theorem": If a prime integer p divides the number of elements in a group then the group must contain an element of order p, that is, a substitution not the identity whose pth power is the identity. For a proof of Cauchy's Theorem see McKay [M1].]

5. Let K be a field containing pth roots of unity (p = prime) and let K' be obtained from K by adjoining a pth root of an element of K. Prove that if $f(x)$ is an irreducible polynomial with coefficients in K then either f is irreducible over K' or it has p irreducible factors of equal degree.

6. Let K be a field, and let p be a prime. Show that if k is an element of K which has no pth root in K then $f(x) = x^p - k$ is irreducible over K. In other words, if $x^p - k$ is reducible then it has at least one linear factor.

7. It was shown in Exercise 28 of the First Set that a polynomial in $x_1, x_2, \ldots, x_n$ can be written in one and only one way as a sum of terms $F(\sigma)x_1^{c_1}x_2^{c_2} \cdots x_n^{c_n}$ where $F(\sigma)$ is symmetric and where the sum has $n!$ terms, one for each choice of the exponents c_i that satisfies $0 \leq c_i \leq n - i$ ($i = 1, 2, \ldots, n$). Use this to prove that the Galois group of the general equation of degree n is the full group of $n!$ substitutions of n objects.

8. Prove the *Eisenstein Irreducibility Criterion*: If $f(x) = x^n + a_1x^{n-1} + a_2x^{n-2} + \cdots + a_n$ is a polynomial with integer coefficients and if there is a prime integer p which divides $a_1, a_2, \ldots, a_n$ but does not divide a_n twice then $f(x)$ is irreducible over $\mathbb{Q}$. [Consider the factors of an assumed factorization mod p.] Apply this to the polynomial $[(x + 1)^p - 1]/x$ to conclude that $(X^p - 1)/(X - 1)$ is irreducible over $\mathbb{Q}$.

9. Prove *Fermat's Theorem* in the form it is used in §70: If p is a prime integer and if $F(X, Y, Z, \ldots)$ is a polynomial in one or more variables $X, Y, Z, \ldots$ with integer coefficients then all coefficients of $F(X, Y, Z, \ldots)^p - F(X^p, Y^p, Z^p, \ldots)$ (which is obviously a polynomial in the same variables with integer coefficients) are divisible by p. [The essential fact was already used in Exercise 8.]

10. Show that Galois' Proposition I applies without change to fields of characteristic p provided $f(x) = 0$ has no multiple roots. Specifically, show that if K is a field obtained from the finite field of integers mod p by the adjunction of a finite number of separable algebraic and/or transcendental elements, if $f(x) = 0$ is an algebraic equation with coefficients in K, and if $f(x)$ and $f'(x)$ have greatest common divisor 1, then there is a group of automorphisms of the splitting field $K(a, b, c, \ldots)$ of f with the property that a polynomial $\Psi(a, b, c, \ldots)$ in the roots $a, b, c, \ldots$ of f is in K if and only if $\Psi(a, b, c, \ldots) = \Psi(Sa, Sb, Sc, \ldots)$ for all S in the group. [If K is infinite, the proof of §32 gives a Galois resolvent and the entire proof is the same. If K is finite, there may be no Galois resolvent, but there is a primitive element for the splitting field, which is sufficient.]

11. Show that the method of §31 of dividing $f(x)$ by its greatest common divisor with $f'(x)$ works — that is, gives a new polynomial with the same roots as f and no multiple roots, so that Proposition I can be applied to the splitting field of f — unless $f(x)$ is divisible by a polynomial in x^p.

12. Show that if K has characteristic p as in Exercise 10, if every element of K has a pth root in K (such a field with characteristic p is called *perfect*), and if $f(x)$ is any polynomial with coefficients in K, then there is a polynomial $g(x)$ with coefficients in K which has distinct roots and the same splitting field as $f(x)$. Thus Proposition I can be applied to the splitting field of $f(x)$. Show that if K is finite then every element of K has a pth root in K.

13. (Artin–Schreier Theorem, Part I.) Let K have characteristic p and let K' be obtained by adjoining a root of $x^p - x - a$ to K for some a in K. Show that the Galois group of any equation $f(x) = 0$ with distinct roots over K' either coincides with its Galois group over K or is a normal subgroup of index p.

14. (Artin–Schreier Theorem, Part II) Let K have characteristic p, let $f(x) = 0$ have coefficients in K and distinct roots, and let H be a subgroup of the Galois group of $f(x) = 0$ over K which is normal of index p. Show that there is an a in K such that adjunction of a root of $x^p - x - a = 0$ to K reduces the Galois group to H. [Difficult.]

15. (Solution of solvable equations.) Let K have characteristic p as above and let $f(x)=0$ (with distinct roots) have a solvable Galois group. Show that a splitting field for f can be constructed by a finite sequence of adjunctions of one of *two* types: (i) solutions of $x^q = a$ where q is prime, $q \neq p$, a is already constructed, and qth roots of unity already adjoined, and (ii) solutions of $x^p - x = a$ where a is already constructed.

16. A field extension is said to be *normal* if any irreducible polynomial with coefficients in the smaller field which has a root in the bigger field splits into linear factors over the bigger field. Show that in the "fundamental theorem of Galois theory" (§63) normal subgroups of the Galois group correspond to normal extensions of K. (In particular, the splitting field, which corresponds to the subgroup containing the identity element alone, is normal.)

17. Show that every normal extension of K of finite degree is the splitting field of some polynomial with coefficients in K. This shows that the "fundamental theorem of Galois theory" can be stated as a theorem applying to normal (separable) algebraic extensions rather than to splitting fields of polynomials.

18. Prove that if K' is a normal extension of K and if the *norm* of an element of K' relative to K is defined as in §60 then *the norm is equal to the product of the conjugates*; that is, $\det(M) = N(a) = \prod Sa$, where M is the $n \times n$ matrix of elements of K which represents multiplication by a relative to some basis of K' over K and where the product is over all elements S of the Galois group of K' over K. [See Exercise 9 of the Seventh Set.]

19. Show that if K' is a normal extension of K and if g is an irreducible polynomial with coefficients in K then the irreducible factors of g over K' all have equal degree.

Memoir on the Conditions for Solvability of Equations by Radicals

by Evariste Galois

Translated by Harold M. Edwards

PRINCIPLES

I shall begin by establishing some definitions and a sequence of lemmas, all of which are known.

Definitions. An equation is said to be reducible if it admits rational divisors; otherwise it is irreducible.

It is necessary to explain what is meant by the word rational, because it will appear frequently.

When the equation has coefficients that are all numeric and rational, this means simply that the equation can be decomposed into factors which have coefficients that are numeric and rational.

But when the coefficients of an equation are not *all* numeric and rational, one must mean by a rational divisor a divisor whose coefficients can be expressed as rational functions of the coefficients of the proposed equation, and, more generally, by a rational quantity a quantity that can be expressed as a rational function of the coefficients of the proposed equation.

More than this: one can agree to regard as rational all rational functions of a certain number of determined quantities, supposed to be known *a priori*. For example, one can choose a particular root of a whole number and regard as rational every rational function of this radical.

When we agree to regard certain quantities as known in this manner, we shall say that we *adjoin* them to the equation to be resolved. We shall say that these quantities are *adjoined* to the equation.

With these conventions, we shall call *rational* any quantity which can be expressed as a rational function of the coefficients of the equation and of a certain number of *adjoined* quantities arbitrarily agreed upon.

When we make use of auxiliary equations, they will be rational if their coefficients are rational in our sense.

One sees, moreover, that the properties and the difficulties of an equation can be altogether different, depending on what quantities are adjoined to it. For example, the adjunction of a quantity can render an irreducible equation reducible.

Thus, when one adjoins to the equation

$$\frac{x^n - 1}{x - 1} = 0, \quad \text{where } n \text{ is prime,}$$

a root of one of Mr. Gauss's auxiliary equations, this equation decomposes into factors, and consequently becomes reducible.

Substitutions are the passage from one permutation to another.

The initial permutation one uses to describe substitutions is entirely arbitrary when one is dealing with functions, because there is no reason, in a function of several letters, for a letter to occupy one position rather than another.

Nonetheless, since one can hardly comprehend the idea of a substitution without that of a permutation, we shall frequently speak of permutations, and we shall consider substitutions only as the passage from one permutation to another.

When we want to group substitutions we shall make them all proceed from the same permutation.

As it is always a question of problems in which the initial distribution of the letters is immaterial, in the groups which we consider one should have the same substitutions no matter which permutation one starts from. Thus if the substitutions S and T are in such a group, one is certain of having the substitution ST.

These are the definitions that we thought we should recall.

LEMMA I. An irreducible equation cannot have a root in common with a rational equation without dividing it.

Because the greatest common divisor of the given irreducible equation and the other equation will also be rational; therefore, etc.

LEMMA II. Given any equation with distinct roots $a, b, c, \ldots$, one can always form a function V of the roots such that no two of the values one obtains by permuting the roots in this function are equal.

For example, one can take

$$V = Aa + Bb + Cc + \ldots,$$

$A, B, C, \ldots$ being suitably chosen whole numbers.

LEMMA III. When the function V is chosen as indicated above, it will have the property that all the roots of the given equation can be expressed as rational functions of V.

In fact,* let

$$V = \phi(a, b, c, d, \ldots),$$

or

$$V - \phi(a, b, c, d, \ldots) = 0.$$

Let us multiply together all the similar equations which one obtains by permuting in these all the letters, leaving just the first one fixed; this will give the following expression:

$$(V - \phi(a, b, c, d, \ldots))(V - \phi(a, c, b, d, \ldots))(V - \phi(a, b, d, f, c, \ldots)) \ldots,$$

which is symmetric in b, c, d, etc., ..., and which can consequently be written as a function of a. We will therefore have an equation of the form

$$F(V, a) = 0.$$

But I say that one can extract from this the value of a. For this it suffices to look for the common solution of this equation and the given one: for one cannot have, for example,

$$F(V, b) = 0$$

unless (this equation having a common factor with the similar equation) one of the functions $\phi(a, \ldots)$ is equal to one of the functions $\phi(b, \ldots)$; which is contrary to the hypothesis.

It therefore follows that a can be expressed as a rational function of V, and it is the same for the other roots.

This proposition† is stated without demonstration by Abel in his posthumous memoir on elliptic functions.‡

---

* We have transcribed word-for-word the demonstration that we gave of this lemma in a memoir presented in 1830. We attach as an historical document the following note which M. Poisson felt he needed to make upon it.

"The demonstration of this lemma is insufficient; however, it is true according to n° 100 of the memoir of Lagrange, Berlin, 1771."

On jugera. (Author's note.)

† It is remarkable that one can conclude from this proposition that every equation depends on an auxiliary equation with the property that all the roots of this new equation are rational functions of one another. For the auxiliary equation for V is of this type.

Moreover, this remark is a mere curiosity; in fact, an equation which has this property is not in general any easier to solve than any other. (Author's note.)

‡ This appears to be a reference to §1 of Chapter 2 of Abel's "Précis d'une théorie des fonctions elliptiques" [A2, p. 547]. Elsewhere [G1, p. 35] Galois says "It would be easy for me to prove that I was unaware even of the name of Abel when I presented my first researches on the theory of equations to the Institute, and that Abel's solution could not have appeared before mine." (Translator's note.)

LEMMA IV. Suppose one has formed the equation for V, and that one has taken one of its irreducible factors, so that V is the root of an irreducible equation. Let V, V', V'', ... be the roots of this irreducible equation. If $a = f(V)$ is one of the roots of the given equation, $f(V')$ will also be a root of the given equation.

In fact, in multiplying together all the factors of the form $V - \phi(a, b, c, \ldots, d)$ in which one applies to the letters all possible permutations, one obtains a rational equation which is necessarily divisible by the equation in question; therefore V' can be obtained by an exchange of letters in the function V. Let $F(V, a) = 0$ be the equation that one obtains in permuting in V all the letters except the first; then one will have $F(V', b) = 0$, where b may be equal to a, but is certainly one of the roots of the given equation. Consequently, just as the given equation and $F(V, a) = 0$ combine to give $a = f(V)$, the given equation and $F(V', b) = 0$ combine to give $b = f(V')$.

With these principles set forth, we shall proceed to the exposition of our theory.

PROPOSITION I

Theorem. Let an equation be given whose m roots are a, b, c, ... There will always be a group of permutations of the letters a, b, c, ... which will have the following property:

1. that each function invariant* under the substitutions of this group will be known rationally;
2. conversely, that every function of the roots which can be determined rationally will be invariant under these substitutions.

(In the case of algebraic equations, this group is none other than the set of all $1 \cdot 2 \cdot 3 \ldots m$ permutations of the m letters, because in this case the symmetric functions are the only ones that can be determined rationally.)
(In the case of the equation $(x^n - 1)/(x - 1) = 0$, if one supposes that

---

* Here we call a function invariant not only if its form is unchanged by the substitutions of the roots, but also if its numerical value does not vary when these substitutions are applied. For example, if $Fx = 0$ is an equation, Fx is a function of the roots which is not changed by any substitution.

When we say that a function is rationally known, we mean that its numerical value can be expressed as a rational function of the coefficients of the equation and the quantities that have been adjoined. (Author's note.)

$a = r, b = r^g, c = r^{g^2}, \ldots$, g being a primitive root, the group of permutations will be simply this one:

$$abcd \ldots k,$$
$$bcd \ldots ka,$$
$$cd \ldots kab,$$
$$\ldots\ldots\ldots$$
$$kabc \ldots i \quad [\text{sic}; i \text{ precedes } k].$$

In this particular case, the number of permutations is equal to the degree of the equation, and the same will be true for equations all of whose roots are rational functions of one another.)

DEMONSTRATION. No matter what the given equation is, one can find a rational function V of the roots such that all the roots are rational functions of V. With such a V, let us consider the irreducible equation of which V is a root (Lemmas III and IV). Let $V, V', V'', \ldots, V^{(n-1)}$ be the roots of this equation.

Let $\phi V, \phi_1 V, \phi_2 V, \ldots, \phi_{m-1} V$ be the roots of the given equation.
Let us write the following n permutations of the roots:

$(V),$	$\phi V,$	$\phi_1 V,$	$\phi_2 V,$	$\ldots,$	$\phi_{m-1} V,$
$(V'),$	$\phi V',$	$\phi_1 V',$	$\phi_2 V',$	$\ldots,$	$\ldots,$
$(V''),$	$\phi V'',$	$\phi_1 V'',$	$\phi_2 V'',$	$\ldots,$	$\ldots,$
$\ldots,$	$\ldots,$	$\ldots,$	$\ldots,$	$\ldots,$	$\ldots,$
$(V^{(n-1)}),$	$\phi V^{(n-1)},$	$\phi_1 V^{(n-1)},$	$\phi_2 V^{(n-1)},$	$\ldots,$	$\phi_{m-1} V^{(n-1)}.$

I say that this group of permutations has the stated property.
In fact:

1. Every function F of the roots invariant under the substitutions of this group can be written as $F = \psi V$, and one will have

$$\psi V = \psi V' = \psi V'' = \cdots = \psi V^{(n-1)}.$$

The value of F can therefore be determined rationally.

2. Conversely, if a function F is determinable rationally, and if one sets $F = \psi V$, one will have

$$\psi V = \psi V' = \psi V'' = \cdots = \psi V^{(n-1)},$$

because the equation for V has no commensurable divisor and V satisfies the rational equation $F = \psi V$, F being a rational quantity. Therefore the function F will necessarily be invariant under the substitutions of the group written above.

Thus, this group has the double property given in the theorem. The theorem is therefore demonstrated.

We will call the group in question the group of the equation.

SCHOLIUM. Clearly in the group of permutations under discussion the disposition of the letters is of no importance, but only the *substitutions* of the letters by which one passes from one permutation to the other.

Thus one can give a first permutation arbitrarily, provided the other permutations are always deduced from it using the same substitutions of the letters. The new group formed in this way will obviously have the same properties as the first, because in the preceding theorem all that matters is the substitutions which one can make in the functions.

SCHOLIUM. The substitutions are independent even of the number of roots.

PROPOSITION II

Theorem. If one adjoins to a given equation the root r of an auxiliary irreducible equation*

(1) one of two things will happen: either the group of the equation will not be changed; or it will be partitioned into p groups, each belonging to the given equation respectively when one adjoins each of the roots of the auxiliary equation;

(2) these groups will have the remarkable property that one will pass from one to the other in applying the same substitution of letters to all the permutations of the first.

(1)† If, after the adjunction of r, the equation for V mentioned above remains irreducible, it is clear that the group of the equation will not be changed. If, on the other hand, it can be reduced, then the equation for V decomposes into p factors, all of the same degree and of the form

$$f(V, r) \times f(V, r') \times f(V, r'') \times \ldots,$$

$r, r', r'', \ldots$ being the other values of r. Thus the group of the given equation also decomposes into groups, each containing the same number of permutations, because each value of V corresponds to a permutation. These groups are, respectively, those of the given equation when one adjoins successively $r, r', r'', \ldots$.

---

* The original version included the words "of prime degree p" which Galois later struck out, perhaps the night before the duel. Thus the letter p, which occurs in the statement of property (1), is intended to be the degree of the auxiliary equation. For the correct statement of the proposition, it should be modified to say that "the group will be partitioned into j 'groups' where j divides p." If p is prime then the partition is into 1 "group" or p. (Translator's note.)

† There is something that needs completing in this demonstration. I haven't the time. (Author's note.)

(2) We saw above that the values of V were all rational functions of one another. In view of this, let V be a root of $f(V, r) = 0$, and $F(V)$ another. It is then clear that if V' is a root of $f(V, r') = 0$, $F(V')$ will be another.*

With this stated, I say that one obtains the group relative to r' by applying the same substitution of letters throughout to the group relative to r.

In fact, if one has, for example,

$$\phi_p F(V) = \phi_n V,$$

one will also have (Lemma I),

$$\phi_p F(V') = \phi_n V'.$$

Therefore, in order to pass from the permutation $(F(V))$ to the permutation $(F(V'))$ one must make the same substitution as one must in order to pass from the permutation (V) to the permutation (V').

The theorem is therefore demonstrated.

(1832. PROPOSITION III

Theorem. If one adjoins to an equation *all* the roots of an auxiliary equation, the groups in Theorem II will have the further property that each group contains the same substitutions.

One will find the proof.†)

PROPOSITION IV

Theorem. If one adjoins to an equation the *numerical* value of a certain function of its roots, the group of the equation will be reduced in such a way as to contain no permutations other than those which leave this function invariant.

In fact, by Proposition I, every known function must be invariant under the permutations of the group of the equation.

---

* Because one will have $f(F(V), r) = $ a function divisible by $f(V, r)$. Therefore (Lemma I) $f(F(V), r') = $ a function divisible by $f(V, r')$. (Author's note.)

† This is a revision made in 1832. The original version was:

PROPOSITION III

THEOREM. If the equation for r has the form $r^p = A$, and if the pth roots of unity have already been adjoined, the p groups of Theorem II will have the further property that the substitutions of letters by which one passes from one permutation to another in each group are the same for all the groups.

In fact, in this case it does not matter which value of r one adjoins to the equation. Consequently, its properties must be the same after the adjunction of any value of r whatever. Thus its group must be the same as far as the substitutions are concerned (Proposition I, Scholium). Therefore, etc. (Translator's note.)

PROPOSITION V

PROBLEM. In which case is an equation solvable by simple radicals?

I shall observe first that in order to solve an equation it is necessary to reduce its group successively until it contains only one permutation. For, when an equation is solved, any function whatever of its roots is known, even when it is not invariant under any permutation.

With this set forth, let us try to find the condition which the group of an equation should satisfy in order that it can be thus reduced by the adjunction of radical quantities.

Let us follow the sequence of possible operations in this solution, considering as distinct operations the extraction of each root of prime degree.

Adjoin to the equation the first radical to be extracted in the solution. One of two things can happen: either by the adjunction of this radical the group of permutations of the equation will be diminished, or, this extraction of a root being only a preparation, the group will remain the same.

In any case, after a certain *finite* number of extractions of roots the group must find itself diminished because otherwise the equation would not be solvable.

If at this point it occurs that there are several ways to diminish the group of the given equation by the simple extraction of a root, it is necessary, in what we are going to say, to consider only a radical of the least possible degree among all the simple radicals which are such that the knowledge of each of them diminishes the group of the equation.

Therefore let p be the prime number which represents this minimum degree such that the extraction of a root of degree p diminishes the group of the equation.

We can always suppose, at least in relation to the group of the equation, that a pth root of unity α is included among the quantities that have already been adjoined to the equation. For, since this expression can be obtained by extractions of roots of degree less than p, its knowledge does not alter in any way the group of the equation.

Consequently, according to Theorems II and III, the group of the equation should decompose into p groups having in relation to one another this double property:
(1) that one passes from one to the other by one single substitution;
(2) that they all contain the same substitutions.

I say that, conversely, if the group of the equation can be partitioned into p groups which have this double property, one can, by a simple extraction of a pth root, and by the adjunction of this pth root, reduce the group of the equation to one of these partial groups.

Let us take, in fact, a function of the roots which is invariant under all substitutions of one of the partial groups, and does not vary [sic] for any other substitution.*

Let θ be this function of the roots.

Let us apply to the function θ one of the substitutions of the total group which it does not have in common with the partial groups. Let θ_1 be the result. Apply the same substitution to θ_1 and let θ_2 be the result, and so forth.

Since p is a prime number, this sequence can end only with the term θ_{p-1}, after which one will have $\theta_p = \theta$, $\theta_{p+1} = \theta_1$, and so forth.

In view of this, it is clear that the function

$$(\theta + \alpha\theta_1 + \alpha^2\theta_2 + \cdots + \alpha^{p-1}\theta_{p-1})^p$$

will be invariant under all the permutations of the total group, and consequently will now be known.

If one extracts the pth root of this function and adjoins it to the equation, then by Proposition IV the group will no longer contain any substitution other than those of the partial groups.

Thus, in order for it to be possible to reduce the group of an equation by simple extraction of a root, the condition stated above is necessary and sufficient.

Let us adjoin to the equation the radical in question; we can now reason with respect to the new group as with respect to the preceding one, and it must be possible to decompose it too in the manner indicated, and so forth, until a group is reached which contains only one permutation.

SCHOLIUM. It is easy to observe this process in the known solution of general equations of the fourth degree. In fact, these equations are resolved by means of an equation of the third degree, which itself requires the extraction of a square root. In the natural sequence of ideas, it is therefore with this square root that one must begin. But in adjoining this square root to the equation of fourth degree, the group of the equation, which contains twenty-four substitutions in all, is decomposed into two which contain only twelve. When the roots are designated by a b c d here is one of these groups:

$abcd$	$acdb$	$adbc$
$badc$	$cabd$	$dacb$
$cdab$	$dbac$	$bcad$
$dcba$	$bdca$	$cbda$

---

* For this it suffices to choose a symmetric function of the various values assumed by a function invariant under no substitutions when it is subjected to the permutations of one of the partial groups. (Author's note.)

Now this group itself splits into three groups, as is indicated in Theorems II and III. Thus, after the extraction of a single radical of third degree just the group

$$abcd$$

$$badc$$

$$cdab$$

$$dcba$$

remains, and this group again splits into two groups

$$abcd \qquad cdab$$

$$badc \qquad dcba.$$

Thus, after a simple extraction of a square root,

$$abcd$$

$$badc$$

remains, which will be resolved, finally, by a simple extraction of a square root.

One obtains in this way either the solution of Descartes or that of Euler. For even though the latter extracts three square roots after the solution of the auxiliary equation of third degree, it is well known that two suffice, because the third can then be derived rationally.

We will now apply this condition to irreducible equations of prime degree.

APPLICATION TO IRREDUCIBLE EQUATIONS OF PRIME DEGREE

PROPOSITION VI

LEMMA. An irreducible equation of prime degree cannot become reducible by the adjunction of a radical. [Sic. Galois evidently means that it cannot become reducible without being solved completely.]

For, if $r, r', r'', \ldots$ are the various values of the radical and if $Fx = 0$ is the given equation, Fx would have to split into factors

$$f(x, r) \times f(x, r') \times \ldots,$$

all of the same degree, which is impossible, at least unless $f(x, r)$ is of the first degree in r. [x]

Thus, an irreducible equation of prime degree cannot become reducible unless its group is reduced to a single permutation.

PROPOSITION VII

PROBLEM. What is the group of an irreducible equation of prime degree n if it is solvable by radicals?

By the preceding proposition, the smallest group possible before the one which contains only a single permutation will contain p permutations. But a group of permutations of a prime number n of letters cannot contain just n permutations unless each of these permutations can be derived from any other by a cyclic substitution of order n (see the memoir of Mr. Cauchy, *Journal de l'Ecole*, **17**).

Thus the next-to-the-last group will be of the form

$$
\begin{array}{llllllll}
x_0 & x_1 & x_2 & x_3 & \cdots & \cdots & \cdots & x_{n-1} \\
x_1 & x_2 & x_3 & x_4 & \cdots & \cdots & x_{n-1} & x_0 \\
x_2 & x_3 & \cdots & \cdots & \cdots & x_{n-1} & x_0 & x_1 \\
\cdots & \cdots & \cdots & \cdots & \cdots & \cdots & \cdots & \cdots \\
x_{n-1} & x_0 & x_1 & \cdots & \cdots & \cdots & \cdots & x_{n-2}
\end{array}
\qquad \text{(G)}
$$

$x_0, x_1, x_2, \ldots, x_{n-1}$ being the roots.

Now the group which immediately precedes this one in the sequence of the decompositions must be made up of a certain number of groups having all the same substitutions as this one. But I observe that these substitutions can be expressed as follows: (Let us set $x_n = x_0$, $x_{n+1} = x_1, \ldots$. It is clear that each of the substitutions of the group (G) can be obtained by putting x_{k+c} in place of x_k throughout, c being a constant.)

Let us consider any one of the groups similar to the group (G). According to Theorem II, it can be obtained by applying one and the same substitution throughout the group, say by putting $x_{f(k)}$ in place of x_k throughout the group (G), f being a certain function.

Since the substitutions of this new group must be the same as those of the group G, one must have

$$f(k + c) = f(k) + C,$$

C being independent of k.

Therefore.

$$f(k + 2c) = f(k) + 2C,$$
$$\cdots\cdots\cdots\cdots\cdots\cdots\cdots\cdots$$
$$f(k + mc) = f(k) + mC.$$

If $c = 1$ and $k = 0$, one finds

$$f(m) = am + b,$$

which is to say

$$fk = ak + b,$$

a and b being constants.

Therefore the group which precedes immediately the group G cannot contain any substitutions other than those of the form

$$x_k \quad x_{ak+b}$$

and consequently can contain no cyclic substitutions other than those of the group G.

One can apply the same argument to this group that was applied to the preceding one, and it follows that the first group in the order of the decompositions, that is, the *actual* group of the equation cannot contain any substitutions other than those of the form

$$x_k \quad x_{ak+b}$$

Therefore "if an irreducible equation of prime degree is solvable by radicals then the group of this equation can contain no substitutions other than those of the form

$$x_k \quad x_{ak+b}$$

a and b being constants."

Conversely, I say that when this condition holds the equation will be solvable by radicals. In fact, consider the functions

$$(x_0 + \alpha x_1 + \alpha^2 x_2 + \cdots + \alpha^{n-1} x_{n-1})^n = X_1,$$

$$(x_0 + \alpha x_a + \alpha^2 x_{2a} + \cdots + \alpha^{n-1} x_{(n-1)a})^n = X_a,$$

$$(x_0 + \alpha x_{a^2} + \alpha^2 x_{2a^2} + \cdots + \alpha^{n-1} x_{(n-1)a^2})^n = X_{a^2},$$

$$\cdots\cdots\cdots\cdots\cdots\cdots\cdots\cdots\cdots\cdots\cdots\cdots\cdots$$

α being an nth root of unity and a a primitive root of n.

It is clear that in this case any function that is unchanged by cyclic substitutions of the quantities X_1, X_a, X_{a^2}, ... will be immediately known. Therefore one can find X_1, X_a, X_{a^2}, ... by the method of Mr. Gauss for binomial equations. Therefore, etc.

Thus, for an irreducible equation of prime degree to be solvable by radicals, it is *necessary* and *sufficient* that every function invariant under the substitutions

$$x_k \quad x_{ak+b}$$

be rationally known.

Thus the function

$$(X_1 - X)(X_a - X)(X_{a^2} - X)\ldots$$

must be known, no matter what X is.

It is therefore *necessary* and *sufficient* that the equation which gives this function of the roots admit, no matter what X is, a rational value.

If the given equation has rational coefficients, the auxiliary equation will also have rational coefficients as well, and it will suffice to determine whether this auxiliary equation of degree $1 \cdot 2 \cdot 3 \ldots (n-2)$ does or does not have a rational root. And one knows how to do this.

This is the method that one must use in practice. But we are going to present the theorem in a different form.

PROPOSITION VIII

Theorem. In order for an irreducible equation of prime degree to be solvable by radicals, it is necessary and sufficient that once any two of the roots are known the others can be deduced from them rationally.

In the first place, it is necessary because, the substitution

$$x_k \quad x_{ak+b}$$

never leaving two letters in the same place, it is clear that when two roots of the equation are adjoined, by Proposition IV, the group is reduced to a single substitution.

In the second place, it is sufficient: because in this case, no substitution of the group can leave two letters in the same place. Consequently the group will contain at the very most $n(n-1)$ permutations. Therefore it will contain only a single cyclic substitution (otherwise it would have at least p^2 [sic; should be n^2] permutations). Therefore each substitution of the group x_k, x_{fk}, must satisfy the condition

$$f(k + c) = fk + C.$$

Therefore, etc.

The theorem is therefore demonstrated.

Example of Theorem VII

Let $n = 5$. The group will be the following one:

> abcde
> bcdea
> cdeab
> deabc
> eabcd
>
> acebd
> cebda
> ebdac
> bdace
> daceb
>
> aedcb
> edcba
> dcbae
> cbaed
> baedc
>
> adbec
> dbeca
> becad
> ecadb
> cadbe

Synopsis

Let K be a field obtained from the rational field $\mathbb{Q}$ by a finite number (possibly none) of simple algebraic and/or transcendental extensions. Let f be a polynomial in one variable with coefficients in K. Then (§§49–61) there is an algebraic extension $L = K(a, b, c, \ldots)$ of K, called the *splitting field* of f over K (unique up to isomorphism), such that $f(x) = k(x - a)(x - b)(x - c)\ldots$, where $k \in K$, where $a, b, c, \ldots$ are the roots of f in L, and where the roots $a, b, c, \ldots$ generate L over K.

Galois seems to have taken for granted—as did his predecessors—the existence of a splitting field, that is, of some universe in which the operations of arithmetic $(+, -, \times, \div)$ could be performed and in which a given equation $f(x) = 0$ had $\deg f$ roots. The first major step in Galois' theory was to give a quite explicit description of the splitting field $L = K(a, b, c, \ldots)$ as a simple algebraic extension $K(t)$ of K. This he did as follows.

Let $n = \deg f$, and let $T = AU + BV + CW + \cdots$ be a linear polynomial in n variables $U, V, W, \ldots$ with coefficients $A, B, C, \ldots$ in the ring of integers $\mathbb{Z}$. Each of the $n!$ ways of substituting the n roots $a, b, c, \ldots \in L$ in place of the variables $U, V, W, \ldots$ of T gives an element of L. Let $t_1, t_2, \ldots, t_{n!}$ be these $n!$ elements of L. When T has the property that $t_1, t_2, \ldots, t_{n!}$ are *distinct* elements of L, T is called a Galois resolvent of $f(x) = 0$ over K. It is easy to show (§32) that if the roots $a, b, c, \ldots \in L$ are distinct then the coefficients $A, B, C, \ldots \in \mathbb{Z}$ of T can be chosen in such a way that T is a Galois resolvent. (If $a, b, c, \ldots$ are not distinct then obviously no choice of $A, B, C, \ldots$ makes T a Galois resolvent. A simple modification of the construction shows, however, that in this case, too, $L = K(t)$ is a simple algebraic extension of K—see §31.) Galois proved (§37) that if T is a Galois resolvent and if t is any one of its $n!$ images in L then t is a primitive element of L over K. Otherwise stated, he proved that every root $a, b, c, \ldots$ of f can be expressed as a rational function of t. (More generally, Galois showed that if T is any *rational function* of the roots $a, b, c, \ldots$ which has $n!$ distinct images in L then $L = K(t)$ when t is any one of these images.)

If T is a Galois resolvent and if $t_1, t_2, \ldots, t_{n!}$ are its $n!$ distinct images in L then the coefficients of the polynomial

$$F(X) = (X - t_1)(X - t_2) \ldots (X - t_{n!})$$

are symmetric functions in the t_i and can therefore be expressed as symmetric polynomials in $a, b, c, \ldots$ with coefficients in $\mathbb{Z}$. Therefore, by the fundamental theorem on symmetric functions (§10), $F(X)$ is a polynomial with known coefficients in K. Let $F = G_1 G_2 \ldots G_v$ be a decomposition of F into factors G_i that are irreducible over K (see §§53–60). The roots of any one of the irreducible factors G_i of F over K are a subset of the roots $t_1, t_2, \ldots, t_{n!}$ of F, and Galois showed, therefore, that any root of any irreducible factor G_i of F is a primitive element of L over K. This shows that the splitting field, assuming it exists, can be described very explicitly as the simple algebraic extension (§34) of K obtained by adjoining a root of any irreducible factor of the known polynomial F. (In particular, the irreducible factors G_i of f all have the same degree, and all of them split into linear factors when one root of any one of them is adjoined to K.)

Galois associated to the equation $f(x) = 0$, with coefficients in the field K (and with distinct roots), a group of substitutions* of the roots $a, b, c, \ldots$ in the following way. Let T be a Galois resolvent, let $F(X) = \prod (X - t_i)$ be the associated polynomial as above, and let G be a factor of F irreducible over K. Then the splitting field $L = K(t)$ is (up to isomorphism) the simple algebraic extension of K obtained by adjoining a root t of G, and $K(t)$ contains $\deg G$ distinct roots of G. The uniqueness assertion of the theorem on simple algebraic extensions implies that t_1 and t_2 are roots of G in $K(t)$ if and only if there is an automorphism of $K(t)$ over K which carries t_1 to t_2. Thus there are $\deg G$ automorphisms of $K(t)$ over K. Since $K(t) = K(a, b, c, \ldots)$, an automorphism of $K(t)$ over K is determined by its effect on the roots $a, b, c, \ldots$. The *Galois group* of $f(x) = 0$ over K is the group of substitutions of $a, b, c, \ldots$ obtained in this way from automorphisms of $K(t)$ over K. It is not difficult to show (§41) that the Galois group is independent of the choice of the Galois resolvent T, the irreducible factor G of F, and the root t of G in L.

If the field K is extended, say to $K' \supset K$, the Galois group of $f(x) = 0$ over K' is in a natural way a *subgroup* of its Galois group over K. To see this,

---

* It would be more natural, in modern terminology, to call a one-to-one mapping of the n roots $a, b, c, \ldots$ to themselves a "permutation" rather than a "substitution." In the book, "substitution" has been used in accord with Galois' usage. Galois appears to have at first used "permutation" to mean both a substitution, in this sense, and an arrangement or ordering of $a, b, c, \ldots$; he then appears to have tried to change his terminology to use "substitution" for substitutions and "permutation" for arrangements. (Some uses of "permutation" for substitutions remained.) Because of this confusion, I have tended to avoid the word "permutation."

note first that the same Galois resolvent T can be used in both cases. Therefore the same $F(X)$ is to be factored. The difference is that the factor $G(X)$ of $F(X)$ that was irreducible over K may not be irreducible over K'. Let $H(X)$ be a factor of $G(X)$ that is irreducible over K'. Since a root t of H is also a root of G, the field $K'(t)$ obtained by adjoining a root t of H to K' *contains* a field $K(t)$ obtained by adjoining a root t of G to K. Therefore, an automorphism of $K'(t)$ over K' restricts to an automorphism of $K(t)$ over K, which shows that any substitution of $a, b, c, \ldots$ obtained from an automorphism of $K'(t)$ over K' can also be obtained from an automorphism of $K(t)$ over K, as was to be shown.

One way to describe the problem of solving an algebraic equation is to say that the field of known quantities K is to be extended until it includes a complete set of roots $a, b, c, \ldots$ of the given equation $f(x) = 0$. If an extension field K' of K does contain n roots $a, b, c, \ldots$ of $f(x) = 0$ then the Galois group of $f(x) = 0$ over K' consists of the identity substitution alone, because automorphisms of $K'(t)$ over K' leave all elements of K' — in particular $a, b, c, \ldots$ — fixed. Conversely, if the Galois group of $f(x) = 0$ over K' consists of the identity substitution alone then deg $H = 1$, and the degree of $K'(t)$ as an extension of K' is 1, which is to say that the splitting field $K'(a, b, c, \ldots)$ over K' is no extension at all, and the roots $a, b, c, \ldots$ are already in K'.

Thus, the problem is to extend K in such a way as to reduce the Galois group of $f(x) = 0$ to the identity substitution alone. If the solution is to be accomplished "by radicals," then the extension of K should be a succession of adjunctions of roots of equations — "pure equations" Gauss called them — of the form $x^m - k = 0$ where m is a positive integer and k is a known quantity. Therefore the problem of solving an equation with radicals turns on the question of determining what reductions of Galois groups can be achieved by adjunctions of this type. Galois gave the answer:

One can assume without loss of generality that m is prime, say $m = p$, and that K contains pth roots of unity (§65). Then, if $f(x) = 0$ is an equation with coefficients in K, and if K' is obtained from K by adjoining the pth root of an element of K, the Galois group of $f(x) = 0$ over K' is either the same as its group over K or it is a normal subgroup of index p (§44). Conversely, if the Galois group of $f(x) = 0$ over K has a normal subgroup of index p then there is an element k of K such that adjunction of a root of $x^p - k = 0$ to K reduces the Galois group to the given subgroup (§46).

This proposition is the main element in the proof that an algebraic equation is solvable by radicals if and only if its Galois group is *solvable*, that is, if and only if the Galois group G has a sequence of subgroups

$$G \supset G_1 \supset G_2 \supset \cdots \supset G_\nu$$

in which each subgroup is a normal subgroup of prime index in its predecessor and in which G_ν contains only the identity substitution (§47).

Other basic facts of Galois theory covered in the book are: The Galois group of the general nth degree equation (with literal coefficients — that is, coefficients that are transcendental over $\mathbb{Q}$) is the full symmetric group of all $n!$ substitutions of the roots (§67). The Galois group of $x^p - 1 = 0$ ($p = $ prime) over $\mathbb{Q}$ is cyclic of order $p - 1$ (§64). More generally, the Galois group of $x^n - 1 = 0$ over $\mathbb{Q}$ is isomorphic to the multiplicative group of invertible integers mod n (§§69-70). An irreducible equation of prime degree p is solvable by radicals if and only if its Galois group is isomorphic to a subgroup of "linear" substitutions $i \mapsto ai + b$ ($a \not\equiv 0 \bmod p$) of the (additive) group of integers mod p (§68). In particular, the general 5th degree equation is not solvable by radicals. Since the symmetric group on five letters is a subgroup of all larger symmetric groups, and since a subgroup of a solvable group is solvable, it follows that the general nth degree equation for $n \geq 5$ is not solvable by radicals.

The Dedekindian tradition, which has dominated algebra for the last century, formulates basic Galois theory somewhat differently. A group is associated not to an equation $f(x) = 0$ with coefficients in K but to a normal extension field $L \supset K$. The group, denoted $\mathrm{Gal}(L/K)$, associated to the normal extension $L \supset K$ is all automorphisms of L which leave elements of K fixed. As was seen above, the Galois group of $f(x) = 0$ over K is isomorphic to $\mathrm{Gal}(L/K)$, where $L = K(a, b, c, \ldots)$ is the splitting field of f over K, and where the isomorphism is given simply by restricting automorphisms in $\mathrm{Gal}(L/K)$ to the n-element subset $\{a, b, c, \ldots\}$ of L.

The advantage of this formulation is that it shows that the group depends (up to isomorphism) only on the splitting field. In particular, the group of an equation $f(x) = 0$ which has multiple roots can be defined in the same way as that of an equation with simple roots, whereas Galois' definition assumes simple roots. The advantage of Galois' original formulation is that it defines the group in a way that makes evident the crucial fact that *extending* the field K *reduces* (or leaves unchanged) the Galois group (§62).

In the modern formulation, one needs to start with a large field Ω containing all extension fields of K that will be considered. The "fundamental theorem of Galois theory" states that there is a one-to-one correspondence between subgroups of $\mathrm{Gal}(\Omega/K)$ and subfields of Ω which contain K(§63). The essence of this theorem is still Galois' Proposition I, which states that an element of a splitting field is invariant under the Galois group only if it is in the ground field — or, in Galois' terms, a rational function $\phi(a, b, c, \ldots)$ of the roots is known if and only if $\phi(a, b, c, \ldots) = \phi(Sa, Sb, Sc, \ldots)$ for all permutations S in the Galois group (§41).

APPENDIX 3
Groups

In modern terminology, a *permutation group* is a subgroup of the group of automorphisms of an n-element set for some positive integer n. (An automorphism of a set is a one-to-one mapping of the set onto itself.) This, by and large, is what Galois meant by a "group," but Galois' conception of a group was strongly influenced by a particular method of *presenting* groups, one that is unfamiliar to modern readers.

Let G be a permutation group operating, for example, on the five letters a, b, c, d, e. Choose any arrangement of the five letters, say the alphabetical order $abcde$. Application of any element of G to this arrangement gives another arrangement. If G has k elements then application of them to $abcde$ gives a list of k arrangements of a, b, c, d, e, including the arrangement $abcde$ corresponding to the identity element of G. Let these k arrangements be listed in k rows with the initial arrangement $abcde$ in the first row. Then, since an element of G is determined by its effect on any one arrangement, the list of k arrangements of a, b, c, d, e determines G as the automorphisms which carry the 1st row of the list to other rows of the list.

It is easy to show that a list of k arrangements of a, b, c, d, e arises in this way from a permutation group G if and only if it has the property that, given any two arrangements in the list, the automorphism which carries the one to the other carries the *first* arrangement in the list to one of the other arrangements in the list. For example, in

$$abcde$$
$$badce$$
$$cdabe$$
$$dcbae$$

the automorphism which carries the third arrangement to the fourth (namely,

$a \leftrightarrow b$, $c \leftrightarrow d$) carries the first to the second. In the same way, the auto-morphism which carries any row to any other carries the 1st row to some row in the list.

A central role in Galois theory is played by the *normal* subgroups of a group. Of course Galois did not use this term, and the concept itself appears only in the unfamiliar and rather confusing statement of his Propositions II and III. An understanding of these propositions requires an examination of the way in which Galois imagined the Galois group of an equation as a "group" of arrangements of the above type.

Let G be the Galois group of $f(x) = 0$ over K. According to the synopsis in Appendix 2, G is obtained by choosing a Galois resolvent $AU + BV + CW + \cdots$, forming the polynomial $F(X) = \prod (X - t_i)$, where $t_1, t_2, \ldots, t_{n!}$ are the $n!$ distinct images of $AU + BV + CW + \ldots$ in $K(a, b, c, \ldots)$ obtained by substituting $a, b, c, \ldots$ in all possible orders for $U, V, W, \ldots$, and factoring $F(X)$ into irreducible factors* $F(X) = G_1(X) \cdot G_2(X) \cdots G_\nu(X)$ over K. The splitting field $K(a, b, c, \ldots)$ is isomorphic to the field $K(t)$ obtained by adjoin-ing to K a root t of any one of the irreducible factors $G_i(X)$ of $F(X)$.

The factorization $F(X) = G_1(X)G_2(X) \cdots G_\nu(X)$ partitions the roots $t_1, t_2, \ldots, t_{n!}$ of $F(X)$ into those that are roots of $G_1(X)$, those that are roots of $G_2(X)$, etc. Since the t_i correspond, by their definition, to arrangements of $a, b, c, \ldots$, this partitions the $n!$ arrangements of $a, b, c, \ldots$ into ν nonover-lapping subsets. (Galois would call these subsets "groups" of arrangements of $a, b, c, \ldots$.) It is virtually the definition of the Galois group (and surely the way Galois thought of it) to say that any one of these subsets presents G as a group of automorphism of the finite set $a, b, c, \ldots$ in the manner spelled out above. In particular, since the Galois group is independent of the choice of the Galois resolvent and the choice of the factor $G_i(X)$, all ν subsets present the same group of automorphisms of $a, b, c, \ldots$.

Now let H be the subgroup of G which is the Galois group of $f(x) = 0$ over an algebraic extension field K' of K. It is natural to assume, because every algebraic extension can be obtained by a succession of simple algebraic extensions (in fact, the theorem of the primitive element states that *one* will suffice), that $K' = K(r)$ is a simple algebraic extension of K obtained by adjoining a root r of irreducible polynomial $g(x)$ with coefficients in K. One obtains presentations of H by factoring $F(X) = H_1(X)H_2(X) \cdots H_\mu(X)$ into polynomials $H_i(X)$ that are irreducible over K' and using this factoriza-tion to partition the $n!$ arrangements of $a, b, c, \ldots$ into μ subsets, each of which is a presentation of the Galois group H of $f(x) = 0$ over K'. Because $K \subset K'$, Galois' Lemma I (§41) implies that each polynomial $G_i(X)$ is equal to a product of a subset of the $H_i(X)$. In this way, the k arrangements of $a, b, c, \ldots$ in any one of the ν presentations of G are partitioned into k/j presentations of H, where j is the number of elements in H. However, this is

---

* The coincidence of the letter G for the Galois group and for the factors $G_i(X)$ of $F(X)$ is un-fortunate but should cause no confusion.

not the partition of the k arrangements in a presentation of G into k/j "groups" to which Galois refers in his Proposition II.

Let $H(X)$ be any irreducible factor of $F(X)$ over K' and let $G(X)$ be the* irreducible factor of $F(X)$ over K which is divisible by $H(X)$. Any element of K', and in particular any coefficient of $H(X)$, can be written as a polynomial in r with coefficients in K. Therefore $H(X)$ can be written in the form $H(X) = H(X, r)$ where $H(X, Y)$ is a polynomial in two variables with coefficients in K. Consider the polynomial

$$H(X, r) \, H(X, r') \, H(X, r'') \ldots, \tag{1}$$

where $r, r', r'', \ldots$ are the roots of $g(x) = 0$. This is a product of polynomials in X with coefficients in the splitting field $K(r, r', r'', \ldots)$ of $g(x)$. However, since it is symmetric in $r, r', r'', \ldots$, its coefficients are in fact in K. (It is, of course, the *norm* of $H(X)$ over K. See §60.)

By assumption, $G(X) = H(X, r)Q(X, r)$ where $Q(X, Y)$ is a polynomial in two variables with coefficients in K. This implies that if $G(X) - H(X, Y)Q(X, Y)$ is written in the form $\Phi_m(Y)X^m + \Phi_{m-1}(Y)X^{m-1} + \cdots + \Phi_0(Y)$, the polynomials $\Phi_i(Y)$ are all divisible by $g(Y)$ (by Galois' Lemma I, because substitution of r for Y must give the zero polynomial). Therefore $G(X) = H(X, r')Q(X, r')$ for any root r' of g in $K(r, r', r'', \ldots)$. Therefore any root of any $H(X, r')$ is a root of $G(X)$. Galois observed that the polynomials $H(X, r')$ *partition* the roots of $G(X)$ because if two of these polynomials have a single root in common they have the *same* roots (and therefore, up to a nonzero multiple, they are the same polynomials). Moreover, this partition of the roots of $G(X)$ has the important property that it is consistent with the action of the Galois group in the sense that if t_0 and t_1 are roots of the same $H(X, r')$ and if t'_0 and t'_1 are their images under any element of the Galois group of $f(x) = 0$ over K (here t_0 and t_1 are identified with the corresponding arrangements of $a, b, c, \ldots$) then t'_0 and t'_1 are roots of the same $H(X, r'')$. This can be proved very simply as follows.

Because every root of $G(X)$ (and in fact every element of $K(a, b, c, \ldots)$) can be expressed rationally in terms of any one root of $G(X)$ (§37), there is a polynomial $\psi(X)$ with coefficients in K such that $t_1 = \psi(t_0)$. Therefore the polynomials $H(\psi(X), r')$ and $H(X, r')$ with coefficients in $K(r')$ have the common root t_0. Since $H(X, r')$ is irreducible over $K(r')$, Galois' Lemma I implies that $H(\psi(X), r') = H(X, r')R(X, r')$. Then, as above, $H(\psi(X), Y) - H(X, Y)R(X, Y) = \Psi_m(Y)X^m + \Psi_{m-1}(Y)X^{m-1} + \cdots + \Psi_0(Y)$ where each $\Psi_i(Y)$ is divisible by $g(Y)$. Therefore $H(\psi(X), r'') = H(X, r'')R(X, r'')$ for all roots r'' of g. Thus if t'_0 is a root of $H(X, r'')$, so is $\psi(t'_0) = t'_1$. In particular, if $t_0 = t'_0$ is a root of both $H(X, r')$ and $H(X, r'')$ and if $t_1 = \psi(t_0)$ is any other

---

* These irreducible polynomials are defined only up to multiplication by nonzero elements of K. It is natural to think of them as having leading coefficient 1 so that they are of the form $\prod (X - t_i)$ where t_i ranges over a subset of $t_1, t_2, \ldots, t_{n!}$.

root of $H(X, r')$ then $t_1 = t'_1 = \psi(t_0)$ is also a root of $H(X, r'')$ which shows that if $H(X, r')$ and $H(X, r'')$ have a root in common their roots coincide.

This partition of the roots of $G(X)$ into subsets consisting of the roots of $H(X, r')$ for various roots r' of $g(x) = 0$ is the "partition of the group into subgroups" that is the subject of Galois' Proposition II. It was just shown to have property (2) of Proposition II. That it has property (1) is the "*quelque chose à compléter*" of which Galois wrote at the last minute. It is easy to fill this small gap provided one makes the obvious modification "the group of the equation will be partitioned into subgroups, the number of which divides the degree of the auxiliary equation" in the statement of property (1):

The partition of the roots of $G(X)$ corresponds to a factorization

$$G(X) = \prod H(X, r^{(i)}), \tag{2}$$

where $r^{(i)}$ ranges over a subset of k/j of the roots $r, r', r'', \ldots$ of $g(x) = 0$. Taking the norm of this equation over K gives $G(X)^{\deg g} = N(X)^{k/j}$ where $N(X)$ is the norm of $H(X, r)$, that is, the polynomial (1) with coefficients in K. Since $G(X)$ is irreducible over K, it follows that $N(X)$ is a power of $G(X)$, say $N(X) = G(X)^m$. Therefore $\deg g = m \cdot (k/j)$, which shows that k/j divides $\deg g$, as desired.

The proof of Proposition III, which Galois said "one will find" is also easy to supply. If all roots of $g(x) = 0$ are adjoined, the resulting extension $K' = K(r, r' r'', \ldots)$ can be obtained by adjoining *one* root s of another auxiliary equation $\bar{g}(x) = 0$, namely an equation $\bar{g}$ of which a Galois resolvent of $g(x) = 0$ is a root. When g is replaced by $\bar{g}$ the factorization (2) takes the form $G(X) = \prod \bar{H}(X, s^{(i)})$, where $s^{(i)}$ runs over some subset of the roots of $\bar{g}$ and $\bar{H}(X, s^{(i)})$ is an irreducible polynomial with coefficients in $K(s^{(i)})$. But the fields $K(s^{(i)})$ are all equal to $K' = K(r, r', r'', \ldots)$. Therefore (2) is a decomposition of $G(X)$ into irreducible polynomials with coefficients in K'. Therefore each $\bar{H}(X, s^{(i)})$ is an irreducible factor of $F(X)$ over K'. Therefore the roots t_i of $F(X)$ that are roots of $\bar{H}(X, s^{(i)})$ give a presentation of the Galois group of $f(x) = 0$ over K', which shows that the groups contain the same substitutions of $a, b, c, \ldots$ for all $s^{(i)}$, as Proposition III states.

If G is a k-element permutation group represented by a list of k arrangements of $a, b, c, \ldots$, and if H is a j-element subgroup of G, then there is a natural way to partition the k arrangements into k/j presentations of H; namely, two arrangements lie in the same subset of the partition if there is an element of H which carries one to the other. At the same time, there is another, less natural, way to partition the k arrangements into k/j subsets; namely, one can pick *one* of the k/j presentations of H mentioned above and apply elements of G to it to obtain the other subsets of the partition. For example, if G is the full six-element symmetric group on three letters and if H is the two-element subgroup generated by $a \leftrightarrow b$ then the first partition is

$$abc \qquad acb \qquad cab$$

$$bac \qquad bca \qquad cba$$

and the second (when the first of these three presentations of H is chosen) is

$$abc \qquad acb \qquad bca$$
$$bac \qquad cab \qquad cba,$$

where the second pair is obtained from the first by applying any element of G which puts b in place of c, and the third pair is obtained from the first by applying any element of G which puts a in place of c. Each pair of the second partition presents a subgroup of G, but these subgroups are conjugate to each other, not equal. The second partition occurs naturally in* Galois' Proposition II because this proposition deals with the *various* subgroups of the Galois group one obtains by adjoining various roots of $g(x) = 0$.

The natural way to define a *normal* subgroup H of G is as a subgroup for which the two methods of partitioning give the same result. Or, more simply, one can say that H is normal in G if the first partition is consistent with the action of G in the sense that application of any element of G to the arrangements in any subset in the first partition gives the arrangements of another subset in the first partition. The reader will have no difficulty in proving that this definition coincides with the usual one.

---

* Galois' description in Proposition II of the partition as a "partition into p groups" shows that he did not always use the word "group" in the modern sense.

Answers to Exercises

First Exercise Set

1. $s^2 - 4p = (x - y)^2$.

2. The product of these two expressions is $\frac{1}{4}[s^2 - (\sqrt{s^2 - 4p})^2] = p$ and their sum is s.

3. The method of §5 leads to $27a^6 + 27(-20)a^3 - 6^3 = 0$ and hence to $a^3 = 10 \pm \sqrt{108}$. Then $b^3 = a^3 - 20 = -10 \pm \sqrt{108}$. The equation $(c + d\sqrt{3})^3 = \pm 10 + \sqrt{108}$ splits into $c(c^2 + 9d^2) = \pm 10$, $d(c^2 + d^2) = 2$. If one tries to find an integral solution, one quickly finds $d = 1$, $c = \pm 1$. Thus $a = 1 + \sqrt{3}$, $b = -1 + \sqrt{3}$, $x = 2$. The factorization $x^3 + 6x - 20 = (x - 2)(x^2 + 2x + 10)$ shows that the product of the other two roots is 10 and their sum is -2. Apply Exercise 2.

4. $27a^6 + 27 \cdot (-2) \cdot a^3 - (-3)^3 = 0$ gives $a^3 = 1$, $a = 1$, $b = -1$, $x = 2$. Since $x^3 - 3x - 2 = (x - 2)(x^2 + 2x + 1)$, the other two roots are both -1. If $a = \omega = \frac{1}{2}[-1 + i\sqrt{3}]$ then $b = -1/\omega = -\omega^2$ and $x = a - b = \omega + \omega^2 = -1$. Similarly $x = -1$ if $a = \bar{\omega} = \omega^2$.

5. From Exercise 3, they are roots of $x^2 + 2x + 10$ and therefore are $-1 \pm 3i$. If $a = 1 + \sqrt{3}$ is replaced by $\omega(1 + \sqrt{3}) = (1/2)(-1-\sqrt{3} + i\sqrt{3} + 3i)$ and b by $(-1 + \sqrt{3})/\omega$ the result is $x = -1 + 3i$. Similarly, $a = \omega^2(1 + \sqrt{3})$ leads to $x = -1 - 3i$.

6. By the binomial theorem, the term in y^{n-1} is $(nC + a)y^{n-1}$. When C is chosen to be $-a/n$ this term is absent.

7. Choose a as in §6. Then $x^4 + px^2 + qx + r = f_+(x)f_-(x)$ where $f_\pm(x) = x^2 + a \pm \sqrt{-p + 2a}\, [x - 2^{-1}q(-p + 2a)^{-1}]$.

8. P is D^3 plus a sum of thirty-three terms, one for each term on the right side of (1)–(19). For example, there are six terms in $g(r)g(s)g(t)$ in which one root occurs to the third power, one to the second, and one to the first, namely, $BC \cdot$ (every r^3s^2t). By (14), this contributes the terms $BCbcd - 3BCd^2$ to P. $P = -d^3 + Bcd^2 - B^2bd^2 + B^3d^2 - Cc^2d + 2Cbd^2 + BCbcd - 3BCd^2 - B^2Ccd - C^2b^2d + 2C^2cd + BC^2bd - C^3d - 3Dbcd + 3Dd^2 + 2DBb^2d + DBcd + 17$ more terms obtained from these by reversing the sign and interchanging B with b, C with c and D with d, that is, $D^3 - bCD^2 + b^2BD^2 - \cdots$.

9. Multiply each term of P by the power of A and the power of a needed to make it of degree 3 in A, B, C, D and in a, b, c, d, that is, $-A^3d^3 + A^2Bcd^2 - AB^2bd^2 + B^3ad^2 - \ldots$. Every term contains A or a, so $P = 0$ if $a = A = 0$. If $a \neq 0$ then $f(x)$ can be factored by §5 as $a(x - r)(x - s)(x - t)$. Then $P = 0$ if and only if $g(r)g(s)g(t) = 0$, which is true if and only if f has a root in common with g.

10. Here A, B, C, D, a, b, c, d are polynomials in y, and consequently P is a polynomial in y. If $(\bar{x}, \bar{y})$ is a root of both equations then $\bar{x}$ is a root of both $f(x, \bar{y})$ and $g(x, \bar{y})$, which shows that $P(\bar{y}) = 0$ unless $A(\bar{y}) = a(\bar{y}) = 0$. This gives a finite set of values of $\bar{y}$ unless $A = a \equiv 0$ or $P \equiv 0$. For each of the $\bar{y}$'s in this finite set, solve the equations $f(x, \bar{y}) = 0$, $g(x, \bar{y}) = 0$ to find all common solutions $\bar{x}$. If $A = a \equiv 0$ then f and g are quadratic at most in x and a similar procedure can be followed using the resultant of two 2nd degree equations. If $P \equiv 0$ without $A \equiv 0$ then there is at least one value of $\bar{x}$ for each value of y, hence ∞ solutions. In this case it is to be expected that f and g have a common factor $h(x, y)$.

11. If m_1, m_2 are monomials, and if m_1 precedes m_2 in lexicographic order, then obviously m_1m_3 precedes m_2m_3. Therefore one of the monomials which goes into fg precedes all the others, namely, the product of the leading terms.

12. The leading terms decrease in lexicographic order and the sequence therefore terminates.

13. For example, if $G = $ every r^3s^3 then $f = \sigma_2^3$. From the expansion $(\alpha + \beta + \gamma)^3 = $ every $\alpha^3 + 3$ every $\alpha^2\beta + 6\alpha\beta\gamma$ with $\alpha = rs, \beta = st, \gamma = tr$, one finds $\sigma_2^3 = $ every $r^3s^3 + 3$ every $r^3s^2t + 6r^2s^2t^2$. The leading term of $G - f = -3$ every $r^3s^2t - 6r^2s^2t^2$ is $-3r^3s^2t$. Thus $g = -3\sigma_1\sigma_2\sigma_3 = -3($ every $r^3s^2t + 3r^2s^2t^2) = -3$ every $r^3s^2t - 9$ every $r^2s^2t^2$. $G - f - g = 3r^2s^2t^2 = 3\sigma_3^2$. Thus $G = \sigma_2^3 - 3\sigma_1\sigma_2\sigma_3 + 3\sigma_3^2 = c^3 - 3bcd + 3d^2$.

14. One way is to set $x = r_n$ in the identity $(x - r_1)(x - r_2)\cdots(x - r_n) = x^n - \sigma_1x^{n-1} + \sigma_2x^{n-2} - \ldots$.

15. If $k = n$, the identity $s_k - \sigma_1s_{k-1} + \cdots = 0$ follows from summation of the identity of Exercise 14 over all variables. The case $k > n$ follows from this one when $k - n$ of the variables are set equal to 0. The case $k < n$ follows from:

Lemma. *A symmetric polynomial of degree $k < n$ in n variables can be written in just one way as every $f(x_1, x_2, \ldots, x_k)$ where f is a symmetric polynomial of degree k.*

17. $x^2 + (a^3 - 3ab)x + b^3 = 0$.

18. $b = b^2, a = 2b - a^2$, hence $x^2 = 0, x^2 - x = 0, x^2 - 2x + 1 = 0, x^2 + x + 1 = 0$ are the only four such equations.

19. By virtue of the lemma in the answer to Exercise 15, one can assume that $n = h + 1$. Then the product is $\sum x_1^{m_1}x_2^{m_2} \ldots x_n^{m_n}$ where $(m_1, m_2, \ldots, m_n)$ ranges over all n-tuples of the form $\tau_1(k_1, k_2, \ldots, k_h, 0) + \tau_2(m, 0, 0, \ldots, 0)$ where τ_1 and τ_2 are permutations that *change* the things they act on. (Thus there are n choices for τ_2 and n times as many terms in the product as in $(k_1, k_2, \ldots, k_h).$) The product will be known once all n-tuples $(m_1, m_2, \ldots, m_n)$ are known in which $m_1 \geq m_2 \geq \cdots \geq m_n$. The rule is a method for counting these n-tuples. [The early editions of [W2] contained an exercise in which this rule was misstated.]

20. Let a symmetric polynomial of the form $(k_1, k_2, \ldots, k_h)$ be called special of length h. The formula for $(k_1)(k_2, k_3, \ldots, k_h)$ gives an expression of $(k_1, k_2, \ldots, k_h)$ as a polynomial with rational coefficients in special polynomials of length $< h$. Repeated use of this expresses any symmetric polynomial in terms of special polynomials of length 1, which are the s_i.

21. $(3, 1) = (3)(1) - (4) = s_3 s_1 - s_4.$ By Newton's theorem, $s_1 = \sigma_1$, $s_4 - \sigma_1 s_3 + \sigma_2 s_2 - \sigma_3 s_1 + 4\sigma_4 = 0$, so $(3, 1) = \sigma_2 s_2 - \sigma_3 s_1 + 4\sigma_4 = \sigma_1^2 \sigma_2 - 2\sigma_2^2 - \sigma_1 \sigma_3 + 4\sigma_4.$ Here $\sigma_1 = -b, \sigma_2 = c, \sigma_3 = -d$, and $\sigma_4 = 0$.

22. Let $F = A_m y_n^m + A_{m-1} y_n^{m-1} + \cdots + A_0$ where the A's are polynomials in $y_1, y_2, \ldots, y_{n-1}$. If $F \neq 0$, then one can divide by a power of y_n, if necessary, to give an F with the same property for which $A_0 \neq 0$. Substitution of $x_n = 0$ shows that A_0 is a polynomial of one fewer variables with the same property as F. Repetition gives a polynomial of *no* variables with these properties, which is impossible.

23. Both sides are polynomials of degree $0 + 1 + 2 + \cdots + (n - 1)$ in the variables which contain the term $a^0 b^1 c^2 d^3 \ldots$ with coefficient $+1$.

Lemma. If $f(x_1, x_2, \ldots, x_n)$ is a polynomial then $f(x_2, x_2, x_3, \ldots, x_n) = 0$ if and only if $f(x_1, x_2, \ldots, x_n) = (x_1 - x_2)g(x_1, x_2, \ldots, x_n)$ for some polynomial g.

For the proof see Exercise 29. Use the lemma to show not only that the polynomial on the left contains all the factors on the right but also that its quotient by any set of these factors still contains the remaining factors.

26. $f'(r_i) = (r_i - r_j)(r_i - r_k).$

27. $$s_1 - (s_2/2) + (s_3/3) - (s_4/4) + \cdots = (\sigma_1 + \sigma_2 + \cdots + \sigma_n)$$
$$-\tfrac{1}{2}(\sigma_1 + \sigma_2 + \cdots + \sigma_n)^2 + \ldots.$$

The terms of degree 4, for example, give $-(s_4/4) = \sigma_4 - \tfrac{1}{2}(2\sigma_1\sigma_3 + \sigma_2^2) + \tfrac{1}{3}(3\sigma_1^2\sigma_2) - \tfrac{1}{4}\sigma_1^4$. The resulting formula for s_k was first given by Waring in the eighteenth century.

28. $z^2 = (\sigma_1^2 - \sigma_2) - \sigma_1 x - \sigma_1 y + xy.$ A polynomial in n variables $x_1, x_2, \ldots, x_n$ can be written in one and only one way in the form $\sum F_{ijk\ldots}(\sigma)x_1^i x_2^j x_3^k \ldots$ where $0 \leq i \leq n - 1$, $0 \leq j \leq n - 2, 0 \leq k \leq n - 3, \ldots$. This is obvious for $n = 1$. Suppose it is known for $n - 1$. A given polynomial can then be written in the form $\sum G_i x_1^i$ where

$$G_i = \sum F(\tau)x_2^\mu x_3^\nu \ldots$$

where the F's are polynomials in the elementary symmetric functions $\tau_1, \tau_2, \ldots, \tau_{n-1}$ in $x_2, x_3, \ldots, x_n$ and where $\mu \leq n - 2, \nu \leq n - 3, \ldots$. The τ's can be expressed in terms of x_1 and σ's, and the degree in x_1 reduced to $< n$ to give an expression in the desired form. It remains to show that an expression of this form $\sum F(\sigma_1, \sigma_2, \ldots, \sigma_n)x_1^i x_2^j x_3^k \ldots$ can be the 0 polynomial in the x's only if all F's are zero. Let σ_n be replaced by $(-1)^n[x_1^n - \sigma_1 x_1^{n-1} + \sigma_2 x_1^{n-2} - \cdots \pm \sigma_{n-1}x_1].$ The result is an expression

$$\sum G(\sigma_1, \sigma_2, \ldots, \sigma_{n-1})x_1^i x_2^j x_3^k \ldots$$

of the same form except that G does not involve σ_n and the degree in x_1 is unrestricted.

Substitution of $x_1 = 0$ and the inductive hypothesis show that the terms free of x_1 are identically 0. Divide by x_1 and repeat this argument to conclude that the G's are identically 0. If the F's are free of σ_n then they are identical to the G's and therefore equal to 0. Otherwise, the terms of highest degree in σ_n that occur among the F's, and, among these, the ones that appear in front of the highest power of x_1, would give a nonzero G in front of the terms in x_1 of highest degree, which is impossible.

29. Suppose it has been shown that $F = (X - r_1) \ldots (X - r_{i-1})Q_i$ where Q_i is a polynomial in $X, r_1, r_2, \ldots, r_n$. Division of polynomials can be used to write $Q_i = (X - r_i)Q_{i+1} + R_{i+1}$ where R_{i+1} is of lower degree in X than $X - r_i$, that is, where X does not occur in R_{i+1}. Substitution of r_i for X in F gives on the one hand 0 (by assumption) and on the other hand $(r_i - r_1)(r_i - r_2) \ldots (r_i - r_{i-1})R_{i+1} = 0$. Since a product of polynomials is 0 only if one of the factors is 0, $R_{i+1} = 0$ and $F = (X - r_1)(X - r_2) \ldots (X - r_i)Q_{i+1}$.

30. Given any solution $x_1, x_2, \ldots, x_n$, substitution of the column vector $y_1, y_2, \ldots, y_n$ for the jth column of the matrix a_{ij} gives a matrix whose determinant is known (it involves a's and y's) and the value of this determinant is $x_j \det(a_{ij})$ (the determinant function is linear in columns, and it is 0 if two columns are equal). This determines x_j unless $\det(a_{ij}) = 0$.

Second Exercise Set

1. Let σ be the cyclic permutation and τ the interchange. Then the six permutations $id, \sigma, \sigma^2, \tau, \sigma\tau, \sigma^2\tau$ are distinct because if $\sigma^i\tau^j = \sigma^m\tau^n$ then $\sigma^{i-m} = \tau^{n-j}$, which implies $\sigma^{i-m} = id = \tau^{n-j}, \sigma^i = \sigma^m, \tau^n = \tau^j$.

2. By direct computation,

$$u = s_3 + 6\sigma_3 + 3\alpha(x^2y + y^2z + z^2x) + 3\alpha^2(xy^2 + yz^2 + zx^2)$$

and v is u with $y \leftrightarrow z$ or, what is the same, with $\alpha \leftrightarrow \alpha^2$. Then $u + v = 2s_3 + 12\sigma_3 - 3$ (every $x^2y) = 2\sigma_1^3 - 9\sigma_1\sigma_2 + 27\sigma_3$, and uv is the cube $(\sigma_1^2 - 3\sigma_2)^3$ of the answer to Exercise 3. (These formulas are in Lagrange [L1], Section 8.)

3. $\sigma_1^2 - 3\sigma_2$.

4. $\pi = t\bar{t} = (x - z)^2 + (y - r)^2 = s_2 - 2(xz + yr)$, which has only three values π_1, π_2, π_3 when x, y, z, r are permuted. Thus $(X - \pi_1)(X - \pi_2)(X - \pi_3)$ has known coefficients.

$$i(t + \bar{t})(t - \bar{t}) = i(2x - 2z)(2iy - 2ir) = -4(xy + zr - xr - yz) = \pm 2\pi_2 \mp 2\pi_3.$$

$$-t^8 + 2t^4\pi^2 - \pi^4 = 4t^4(\pi_2 - \pi_3)^2 = 4t^4[(\pi_2 + \pi_3)^2 - 4\pi_2\pi_3]$$

and $\pi_2 + \pi_3$ and $\pi_2\pi_3$ can be expressed in terms of $\pi = \pi_1$ and known quantities. This gives rise to an explicit factorization of the equation for t into twenty-four linear factors. $x = (\sigma_1 + t + \bar{t} + u)/4$ where $u = x - y + z - r$. If t_2 is a value of t other than $i^k t$ or $i^k\bar{t}$ then $\pm \text{Re}(t_2) \pm \text{Im}(t_2)$ has four values, two of which are among the four values $\pm \text{Re}(t) \pm \text{Im}(t)$ and one of which is u. This gives two values for x, one of which is a solution. I do not know how to determine t_2 so that $u = \text{Re}(t_2) + \text{Im}(t_2)$.

5. Straightforward multiplication gives $t_1 t_2 t_3 = $ every $x^3 - $ every $x^2y + 2$ every xyz. In the notation of Exercise 19 of the First Set, every $x^2y = (2, 1) = (2)(1) - (3) = s_2 s_1 - s_3$. Thus $t_1 t_2 t_3 = 2s_3 - s_1 s_2 + 2\sigma_3 = \sigma_1^3 - 4\sigma_1\sigma_2 + 8\sigma_3$.

7. The main fact is that $1 + \alpha + \alpha^2 + \alpha^3 + \alpha^4 = (1 - \alpha^5)/(1 - \alpha) = 0$. There are many ways to choose t_1, t_2, t_3, t_4; for example, leave x_1 alone and cyclically permute x_2, x_3, x_4, x_5.

8. If no three of x, y, z, r are equal then they can be reordered to make $x \neq y, z \neq r$. Then $x - y + z - t \neq \pm(-x + y + z - t)$.

Third Exercise Set

4. Let $x^j = 1$ and $y^k = 1$ be primitive roots of unity. If $(xy)^m = 1$ then $(xy)^{mj} = 1$, from which $y^{mj} = 1$. It follows that mj is divisible by k. Since j and k are relatively prime, m is divisible by k, say $m = qk$. Then $(xy)^{qk} = x^{qk} = 1$, from which j divides q. Thus jk divides m.

5. Changing α to α^2 in the formula gives

$$\alpha^2 = \tfrac{1}{10}[\beta^{-1}t + (t_2 t^8)/(\beta^{-1}t)^8 + (t_3 t^7)/(\beta^{-1}t)^7 + \ldots].$$

Thus, changing t to $\beta^{-1}t$ in the formula gives α^2. Similarly, changing α to α^{2^k} in the formula shows that this 11th root of unity is given by the formula when t is replaced by $\beta^{-k}t$. Thus each 11th root $\alpha, \alpha^2, \alpha^3, \alpha^4, \ldots, \alpha^{10}$ can be obtained by using a suitable root $\beta^{-k}t$ of $x^{10} = t^{10}$ and these roots $\beta^{-k}t$ all give 11th roots of unity.

7. See [E1], Appendix A.2. Here is an outline of Gauss's other proof. Let $\psi(d)$ be the number of integers mod p whose orders mod p are exactly d. If any integer has order d mod p then all d of its powers satisfy $x^d - 1 \equiv 0$ mod p. Show that this congruence has at most d roots. Thus if $\psi(d) > 0$ then $x^d \equiv 1$ mod p has d roots. Show that $\phi(d)$ of them have order d, where $\phi(d)$ is the number of integers less than d relatively prime to d. Thus $p - 1 = \sum_{d|p-1} \psi(d) \leq \sum_{d|p-1} \phi(d)$. Prove that $\sum_{d|p-1} \phi(d) = p - 1$ and that $\phi(p - 1) > 0$. Conclude that $\psi(p - 1) > 0$, as desired. Note that this proof also shows, more generally, that *the multiplicative group of any finite field is cyclic*.

8. Use the argument of §24 with $\beta = 1$.

9. $K_1 \subset K_2 \subset K_6 \subset K_{18}$, or $K_1 \subset K_3 \subset K_9 \subset K_{18}$, or $K_1 \subset K_3 \subset K_6 \subset K_{18}$. In general, what is sought is a sequence $1 < d_1 < d_2 < \cdots < d_m = p - 1$, where $d_i | d_{i+1}$. Then $p - 1 = (d_m/d_{m-1})(d_{m-1}/d_{m-2}) \ldots (d_1/1)$ is a product of m factors > 1.

10.
$$t_1 = \alpha - \alpha^4,$$
$$t_1^2 = \alpha^2 - 2 + \alpha^3 = -3 - \gamma_2,$$
$$t_2 = \gamma_2 - S(\gamma_2) = \alpha + \alpha^4 - \alpha^2 - \alpha^3$$

and by direct computation $t_2^2 = 5$. Then $2\gamma_2 = t_2 + \gamma_2 + S(\gamma_2) = t_2 - 1$ and $2\alpha = t_1 + \alpha + \alpha^4 = t_1 + \gamma_2$ gives the final formulas.

11. Set $t_i = (\alpha + \alpha^{-1}) + \omega^i(\alpha^2 + \alpha^{-2}) + \omega^{2i}(\alpha^3 + \alpha^{-3})$ and compute $t_1 t_2$, t_1^2/t_2. $\gamma = (-1 + t + 7t^{-1})/3$ where t is the cube root of $7(2 + 3\omega)$.

12. If u is any element of K_2 then $u = a\theta + bS\theta$ where $\theta = \alpha + S^2\alpha + S^4\alpha + \cdots + S^{14}\alpha$ and a and b are rational. To evaluate u by the method of §25 one uses the resolvent $u - Su = (a - b)(\theta - S\theta)$. Let $t = \theta - S\theta$. Straightforward computation gives $t^2 = 17$. The sign of t is $+$ because $t = c_1 - c_3 + c_8 - c_7 + c_4 - c_5 + c_2 - c_6$ where $c_k = 2\cos(2\pi k/17)$; the only negative quantities in this sum are $-c_3$ and c_8, which are more than compensated for by c_2 and $c_1 - c_7$, as is clear from a diagram and can be verified

easily. Therefore $t = \sqrt{17}$, and $u - Su = (a - b)\sqrt{17}$. Since $u + Su = -a - b$, it follows that $u = \frac{1}{2}[-a - b + (a - b)\sqrt{17}]$. Otherwise stated, $u = a\theta + bS\theta$ can be found using $\theta = (-1 + \sqrt{17})/2$, $S\theta = (-1 - \sqrt{17})/2$. Similarly, if $u \in K_4$ then $u = a\eta + bS\eta + cS^2\eta + dS^3\eta$ where $\eta = \alpha + S^4\alpha + S^8\alpha + S^{12}\alpha$ and a, b, c, and d are rational. It will suffice, therefore, to find η, $S\eta$, $S^2\eta$, $S^3\eta$. Since $\eta + S^2\eta = \theta$ and $S\eta + S^3\eta = S\theta$ are known, it will suffice to find η and $S\eta$. The corresponding resolvents are $s = \eta - S^2\eta = c_1 - c_8 + c_4 - c_2$ and $Ss = c_3 - c_7 + c_5 - c_6$, both of which are positive. Computation gives $s^2 = 8 - \theta = (17 - \sqrt{17})/2$, from which $(Ss)^2 = S(8 - \theta) = (17 + \sqrt{17})/2$, and all elements of K_4 can be evaluated. Finally, let $\zeta = \alpha + \alpha^{-1} = \alpha + S^8\alpha$. The resolvent $\zeta - S^4\zeta = c_1 - c_4$ is positive, and its square can be computed. The desired quantity $\zeta/2$ can then be found.

13. Let

$$\zeta = \alpha + S^8\alpha, \eta = \zeta + S^4\zeta, \theta = \eta + S^2\eta, t = \theta - S\theta = 2\theta + 1, s = \eta - S^2\eta$$

$$= 2\eta - \theta, r = \zeta - S^4\zeta = 2\zeta - \eta.$$

The desired quantity is $\zeta/2 = (4r + 2s + t - 1)/16$. Explicit computation gives $t^2 = 17$, $s^2 = (17 - t)/2$, $r^2 = S^2\eta + 4 - 2S\eta$. From $\eta = (2s + t - 1)/4$, the value $S^2\eta = (-2s + t - 1)/4$ follows. The value of $S\eta$ is less obvious. It suffices to find Ss. From $(s \cdot Ss)^2 = s^2 \cdot S(s^2) = (17 - t)(17 + t)/4 = 4 \cdot 17$ we have $s \cdot Ss = \pm 2t$. Examination of the coefficient of α in $s \cdot Ss$ gives $s \cdot Ss = 2t$. This gives the formula

$$\frac{\zeta}{2} = -\frac{1}{16} + \frac{t}{16} + \frac{s}{8} \pm \frac{1}{8}\sqrt{17 + 3t - 2s - 4Ss}.$$

Application of S^4 merely changes the sign of the radical. Therefore the eight desired values are given by the four formulas obtained from this one by applying S^0, S^1, S^2, S^3 to s and t in this one and evaluating Ss using $Ss = 2t/s$. These values are in the field $\mathbb{Q}(t, s, q)$ where $t^2 = 17$, $s^2 = (17 - t)/2$, and $q^2 = 17 + 3t - 2s - 8ts^{-1}$. They can be realized as real numbers by setting $t = \sqrt{17}$, $s = \frac{1}{2}\sqrt{34 - 2\sqrt{17}}$, and evaluating S^jt, S^js accordingly.

15. Let θ be the p^{n-1}st root of α. Then θ to the power p^n is 1. If $\theta^k = 1$ then $\theta^d = 1$ where d is the greatest common divisor of k and p^n. Then, if $d \neq p^n$, α is a power of $\theta^d = 1$, which is impossible. Therefore $p^n | k$.

Fourth Exercise Set

1. Determine i by $\psi t = \phi_i t$ and set $g = \phi_i^{-1}\psi$.

2. First prove that $(f + g)' = f' + g'$ and $(fg)' = f'g + fg'$. Clearly the derivative of a constant is 0 and the derivative of $f(x) = x$ is 1. Prove by induction that the derivative of $f(x) = (x - a)^n$ is $n(x - a)^{n-1}$ for $n = 2, 3, \ldots$. If $f(x)$ is a given polynomial and if a is one of its roots then $f(x) = (x - a)^n q(x)$ for some integer $n > 0$, where $q(x)$ is a polynomial not divisible by $x - a$ or, what is the same, of which a is not a root. (See note on the Remainder Theorem in §37.) Then $(x - a)^{n-1}$ divides both terms of f', while $(x - a)^n$ divides one term but not the other, so $(x - a)^{n-1}$ divides $d(x)$ but $(x - a)^n$ does not. Let $Q(x) = f(x)/d(x)$. Then $(x - a)^n$ divides $f(x) = d(x)Q(x)$ but not $d(x)$, from which one can deduce that a is a root of $Q(x)$. If $(x - a)^2$ divided $Q(x)$ then $(x - a)^{n+1}$ would divide $f(x)$.

3. First consider polynomials in one variable $F(A)$. A root of a polynomial implies a factor. Therefore $F(x_k) \neq 0$ for some $k = 1, 2, \ldots, \deg F + 1$. Once the theorem is known for polynomials in $n - 1$ variables, and for one variable, it is easy to deduce for n variables.

4. In lexicographic order, the leading term of the product is the product of the leading terms.

5. Adjoin a root i of $x^2 + 1$.

6. The method of §36 calls for finding a g.c.d. of $2x + 3$ and $x^2 - 2$. From $x^2 - 2 = (x/2 - \frac{3}{4})(2x + 3) + \frac{1}{4}$, this is achieved by $1 = 4(x^2 - 2) - (2x - 3)(2x + 3)$, which gives $3 - 2\sqrt{2}$ as the inverse. Alternatively, one can "rationalize the denominator."

7. Equality of any two of $a - b, a - c, b - a, b - c, c - a, c - b$ implies either the equality of two of a, b, c, or implies that one root times 3 is $a + b + c$ and therefore that this root is rational.

8.

$$(X - (a - b))(X - (b - c))(X - (c - a)) = X^3 + [(a - b)(b - c) + (b - c)(c - a)$$
$$+ (c - a)(a - b)]X - (a - b)(b - c)(c - a) = X^3 + 3pX + P$$

where P is so defined. Then $F(X) = (X^3 + 3pX + P)(X^3 + 3pX - P) = X^6 + 6pX^4 + 9p^2X^2 - P^2$ and $P^2 = (a - b)^2(b - a)^2(c - a)^2 = -4p^3 - 27q^2$. If P^2 has a square root in K, this gives a factorization of $F(X)$.

9. $F(t, a) = 0$ where $F(X, Y)$ is found by expressing $(X - (Y - b))(X - (Y - c))$ in terms of a and putting Y in place of a. This gives $(X - Y)^2 + (b + c)(X - Y) + bc = (X - Y)^2 - a(X - Y) + (ab + bc + ca) - a(b + c)$, hence $F(X, Y) = (X - Y)^2 - Y(X - Y) + p + Y^2 = X^2 - 3XY + 3Y^2 + p$. Thus $3a^2 - 3ta + t^2 + p = 0$. This combines with $a^3 + pa + q = 0$ to give $a(2p + 2t^2)/3 + (3q - t^3 - pt)/3 = 0$. Similar computations apply to b and c to give

$$a = \frac{t}{2} - \frac{3}{2}\left(\frac{q}{p + t^2}\right), \qquad b = -\frac{t}{2} - \frac{3}{2}\left(\frac{q}{p + t^2}\right), \qquad c = \frac{3q}{t^2 + p}.$$

Of course the other two can be deduced from any one of these without further recourse to Galois' method. Computation of c^3 entails the evaluation $(t^2 + p)^3 = t^6 + 3pt^4 + 3p^2t^2 + p^3 = -3pt^4 - 6p^2t^2 - 3p^3 - 27q^2$ (because $t^6 + 6pt^4 + 9p^2t^2 + 4p^3 + 27q^2 = 0$) from which $c^3 + pc + q = 0$ follows easily.

10. $c = t^4/9r$. $a/c = (t^3/6q) - \frac{1}{2}.(t^3/3q)^2 = -27 r^2/9r^2 = -3$.

11. $t^6 + 6pt^4 + 9p^2t^2 + 4p^3 + 27q^2 = (t^2 + p)(t^4 + 5pt^2 + 4p^2) + 27q^2$.

Therefore $(t^2 + p)^{-1} = -(t^4 + 5pt^2 + 4p^2)/27q^2$, $c = -(t^4 + 5pt^2 + 4p^2)/9q$, $a = (t - c)/2$, $b = (-t - c)/2$. (Note that $q \neq 0$.)

12. In Exercise 9, $c = 3q(t^2 + p)^{-1}$ is derived from $t^2 + 3c^2 + 4p = 0$ and $c^3 + pc + q = 0$. Conversely, if c is defined by this formula, where $t^6 + 6pt^4 + 9p^2t^2 + 4p^3 + 27q^2 = 0$ then an easy computation gives $(t^2 + 3c^2 + 4p)(t^2 + p)^2 = 0$. Therefore $(t^2 = -p$ would imply $q = 0)$ $t^2 + 3c^2 + 4p = 0$. Then $c^3 + pc + q = 0$ can be derived either by reversing the steps in the derivation of $c = 3q(t^2 + p)^{-1}$ or as in Exercise

9. Set $a = (t - c)/2$, $b = (-t - c)/2$. Then $(x - a)(x - b) = x^2 + cx + (c^2 - t^2)/4 = x^2 + cx + c^2 + p$, so that $(x - a)(x - b)(x - c) = x^3 - c^3 + p(x - c) = x^3 + px + q$, as desired.

13. Let $f_r(X), f_s(X), f_t(X), \ldots$ be polynomials with coefficients in K of which $r, s, t, \ldots$, respectively, are roots. Set $f = f_r f_s f_t \ldots$. Then f is a polynomial with coefficients in K and $r, s, t, \ldots$ is subset of its roots.

14. Assume without loss of generality that the polynomial in question has no multiple roots. For simplicity of notation, let a, b, c be the given subset of $a, b, c, d, \ldots$. Use the argument of §32 to find a rational function ϕ of three variables such that the $n(n - 1)(n - 2)$ values $\phi(a, b, c)$, $\phi(a, b, d)$, $\phi(a, c, d), \ldots$ are all distinct. As before, one can assume $\phi(x, y, z)$ has the special form $Ax + By + Cz$ where A, B, C are integers. Set $u = \phi(a, b, c)$ and $F(X, Y) = (X - \phi(Y, b, c))(X - \phi(Y, b, d)) \ldots$ where the product has $(n - 1)(n - 2)$ terms, where the symmetric polynomials in $b, c, d, \ldots$ are expressed in terms of a, and a is replaced by Y. Then a is the only common root of $f(Y)$, and $F(u, Y)$, as before, and this shows that a can be expressed rationally in terms of u. Similarly, b and c can be expressed rationally in terms of u.

15. See Exercise 28 of the First Set.

17. Explicit computation of $\mathscr{D}$ (see §32) gives $D(A - B)^2$ where D is the discriminant.

18. Say $a(X) = a_k X^k + \ldots$ and $b(X) = b_v X^v + \cdots$. If $k < v$ set $q(X) = 0$, $r(X) = a(X)$. Otherwise let $c(X) = a(X) - (a_k/b_v)b(X)X^{k-v}$. By the induction hypothesis $c(X) = Q(X)b(X) + R(X)$. Set $q(X) = Q(X) + (a_k/b_v)X^{k-v}$, $r(X) = R(X)$. If $0 = q(X)b(X) + r(X)$ then $q(X) = 0$ because otherwise deg qb would be greater than deg$(-r)$.

Fifth Exercise Set

1. Let the substitutions $S_1, S_2, \ldots, S_m$ form a group. Then, by definition, there are arrangements $A_1, A_2, \ldots, A_m$ which present the group. This implies that $S_i(A_j)$ is of the form A_k for all i, j. Thus $S_r(S_i(A_j)) = S_r(A_k) = A_s$. Since $S_r \circ S_i$ carries A_j to A_s it is one of the substitutions $S_1, S_2, \ldots, S_m$ and they are closed under composition. Now suppose $S_1, S_2, \ldots, S_m$ are closed under composition. Because all powers of S_1 are in the finite set $S_1, S_2, \ldots, S_m$, there must be integers $i > j > 0$ for which $S_1^i = S_1^j$. Then Then $S_1^{i-j} \circ S_1^j = S_1^j$ and S_1^{i-j} is the identity substitution. If $i = j + 1$ then S_1 is the identity. Otherwise S_1^{i-j-1} is a substitution in the set and is inverse to S_1. In any case, then, the identity is among the substitutions $S_1, S_2, \ldots, S_m$ and every substitution in the set has an inverse in the set. Let A be any arrangement and let $A_i = S_i(A)$. The substitution which carries the ith arrangement to the jth is then $S_j S_i^{-1}$. It is to be shown that for any k the substitutions $S_j S_i^{-1}$ for $j = 1, 2, \ldots, m$ coincide with the substitutions $S_r S_k^{-1}$ for $r = 1, 2, \ldots, m$. That each j corresponds to one and only one r follows from $S_j S_i^{-1} = S_r S_k^{-1}$, $S_r = S_j S_1^{-1} S_k$ and $S_j = S_r S_k^{-1} S_1$.

2. Let the arrangements $A_1, A_2, \ldots, A_M$ be a presentation of the group $G = \{S_1, S_2, \ldots, S_M\}$, where it is assumed that no two A's are equal, so that one and only one S is the identity. Let $H = \{S_1, S_2, \ldots, S_m\}$ be a subgroup of G. Renumber the A's so that $A_1, A_2, \ldots, A_m$ are the arrangements to which elements of H carry A_1. (A_1 is one of these because H contains the identity.) It was seen in the preceding answer that if A is any arrangement and if $\{S_1, S_2, \ldots, S_m\}$ is any group then $S_1(A), S_2(A), \ldots, S_m(A)$ is a presentation of the group. Thus $A_1, A_2, \ldots, A_m$ is a presentation of H. By the same

token if $M > m$ then $S_1(A_{m+1})$, $S_2(A_{m+1})$, ..., $S_m(A_{m+1})$ is a presentation of H. It is clear from the definition that if two presentations have an arrangement in common then they contain exactly the same arrangements. Therefore the second presentation is disjoint from the first one and A_{m+1}, A_{m+2}, ..., A_M can be renumbered to make A_{m+1}, A_{m+2}, ..., A_{2m} the arrangements of the second presentation of H. Similarly, if $M > 2m$ then A_{2m+1}, ..., A_M can be renumbered so that A_{2m+1}, A_{2m+2}, ..., A_{3m} are a presentation of H. This process terminates with a renumbering of the original presentation of G in which. for each j, $0 < j < M/m$, A_{jm+1}, A_{jm+2}, ..., A_{jm+m} is a presentation of H.

3. Let A_1, A_2, ..., A_M be a presentation of G. It was shown in the preceding exercise that the A's can be renumbered so that A_{jm+1}, A_{jm+2}, ..., A_{jm+m} is a presentation of H for $j = 0, 1, ..., (M/m) - 1$. Suppose this presentation presents H as a normal subgroup. Then for $j = 0. 1, ..., (M/m) - 1$ there is a substitution T_j such that $T_j(A_1)$, $T_j(A_2)$, ..., $T_j(A_m)$ and A_{jm+1}, A_{jm+2}, ..., A_{jm+m} are the same presentation of H, that is, contain the same arrangements, though possibly in a different order. If T is any element of G then $T(A_1) = A_{jm+k}$ for some $j = 0, 1, ..., (M/m) - 1$ and $k = 1, 2, ..., m$. Thus $T(A_1) = T_j(A_r)$ for some $r = 1, 2, ..., m$. Since $A_r = S_r(A_1)$ where $S_r \in H$, $T = T_j S_r$ for $S_r \in H$. It follows that for $s = 1, 2, ..., m$, $T(A_s) = T_j S_r(A_s) = T_j S_r S_s(A_1) = T_j(A_t) = A_{jm+u}$ for some $u = 1, 2, ..., m$. The arrangements $T(A_1)$, $T(A_2)$, ..., $T(A_m)$ are therefore included among A_{jm+1}, A_{jm+2}, ..., A_{jm+m}. Since there are m distinct arrangements in both cases, they coincide and $T(A_1)$, $T(A_2)$, ..., $T(A_m)$ is a presentation of H. Therefore, for any $S \in H$, $ST(A_1)$ is of the form $T(A_i)$ for $i \leq m$. Thus $T^{-1}S T$ carries A_1 to A_i, which shows that $T^{-1}S T$ is in H, as was to be shown. Now suppose that $T^{-1}S T$ is in H whenever S is in H and T in G. It is to be shown that any one of the presentations A_{jm+1}, A_{jm+2}, ..., A_{jm+m} can be obtained from any other by application of a single substitution T. It will suffices to find, for each j, a T_j such that $T_j(A_1)$, $T_j(A_2)$, ..., $T_j(A_m)$ contains the same arrangements as A_{jm+1}, A_{jm+2}, ..., A_{jm+m}. For this, let T_j be the substitution which carries A_1 to A_{jm+1}. For $k = 1, 2, ..., m$, $T(A_k) = TS_k(A_1) = TS_k T^{-1}(A_{jm+1}) = S'(A_{jm+1})$ where $S_k \in H$ and therefore $S' \in H$. It follows that $T(A_k)$ is one of the arrangements A_{jm+1}, A_{jm+2}, ..., A_{jm+m}. Therefore these presentations coincide, as was to be shown.

4. Let g be a factor of f. A factorization $g = g_1 g_2$ of g into polynomials of lesser degree partitions the roots of g into two disjoint (because f has no multiple roots) nonempty subsets, namely, the roots of g_1 and the roots of g_2. Since $g_1(a) = 0$ implies $g_1(S(a)) = 0$ for all substitutions S in the Galois group (S gives an automorphism which leaves the coefficients of g_1 fixed because they are in K) no S can carry a to a root of g_2 and the action is not transitive. Conversely, if the action is not transitive, let a be a root such that not every root is of the form $S(a)$. Let $h(X) = \prod (X - S(a))$ where $S(a)$ ranges over all *distinct* roots of f which have the form $S(a)$. Then $h(X)$ divides $g(X)$ and it has positive degree less than deg g. When it is multiplied out, it has coefficients in $K(a, b, c, ...)$ that are invariant under the Galois group (a substitution of the Galois group merely permutes the factors of h), which shows, by Proposition I, that they are in K.

5. Enumeration of the subgroups of the full group (there are six altogether) shows that only these two act transitively. By Exercise 8 of the Fourth Set, $F(X) = (X^3 + 3pX + P)$ $(X^3 + 3pX - P)$. Thus, if $P \in K$ then F factors and the Galois group has three elements. Conversely, if the Galois group has three elements then $P = (a - b)(b - c)(c - a) \in K$ by Proposition I.

6. By Exercises 8 and 11 of the Fourth Set, $c = -(t^4 + 5pt^2 + 4p^2)/9q$ is a root of $x^3 + px + q$ when t is a root of $X^6 + 6pX^4 + 9p^2X^2 + 4p^3 + 27q^2$. If $t^3 + 3pt + P = 0$ then $c = (Pt - 2pt^2 - 4p^2)/9q$. The other roots of $x^3 + px + q = 0$ are represented as elements of $K(t)$ by $a = (t - c)/2$, $b = (-t - c)/2$. Define $t' = b - c = (-t - 3c)/2$ and $t'' = c - a = (-t + 3c)/2$. Then, because $t^2 + 3c^2 + 4p = 0$ (Exercise 12, Fourth Set), $(X - t')(X - t'') = X^2 + tX + t^2 + 3p$, $(X - t)(X - t')(X - t'') = X^3 + 3pX + P$. This shows that t', t'' are conjugates of t in $K(t)$. Replacement of t by t' in $a = \phi_a(t)$, $b = \phi_b(t)$, $c = \phi_c(t)$ gives roots a', b', c' of $x^3 + px + q = 0$. Since $t = a - b$, $t' = a' - b'$. Therefore, because the six versions of $a - b$ are all distinct and because $t' = b - c$, $a' = b$ and $b' = c$. Therefore $c' = a$. Also $t'' = c - a = b' - c'$ is the image of t'. Repetition of $t \mapsto t'$ carries t to t'', a to c, b to a, and c to b. A third application of $t \mapsto t'$ gives the identity of $K(t)$.

7. The expressions of a, b, c as polynomials in t were found in Exercise 11, Fourth Set. The preceding exercise gives $(X - t)(X - t')(X - t'') = X^3 + 3pX + P$ where $P = -t^3 - 3pt$. Thus $(X - t)(X - t')(X - t'')(X + t)(X + t')(X + t'') = (X^3 + 3pX)^2 - P^2 = F(X)$.

8. If the expressions of $1, a, a^2$ in terms of $1, t, t^2$ can not be solved for $1, t, t^2$ in terms of $1, a, a^2$ then $1, a, a^2$ satisfy a linear relation $\alpha \cdot 1 + \beta \cdot a + \gamma \cdot a^2 = 0$ (with α, β, γ in K) contrary to the assumption that a is the root of an irreducible cubic. $b, c = -(a/2) \pm (6pa^2 - 9qa + 4p^2)/2P$. In the first case $b, c = a^2 - 2, -a^2 - a + 2$. In the second, $b, c = (-1 \pm \sqrt{-3})a/2$. In the second case it is assumed, of course, that $\sqrt{-3} \in K$ so that the discriminant is a square.

9. The roots of F in $K(a, b, c, \ldots)$ all have the form $AS(a) + BS(b) + CS(c) + \ldots = t_S$ where S is a substitution of the roots $a, b, c, \ldots$. Therefore t_S is a Galois resolvent. The degree of the irreducible factor (over K) of F of which t_S is a root is the number of substitutions in the Galois group, and this is independent of t_S.

10. It was shown in §42 that the Galois group contains no other substitutions than the ones presented in (1). By Exercise 4, the Galois group must contain at least one substitution which carries a_0 to a_i for each $i = 0, 1, 2, \ldots, p - 2$. Therefore all $p - 1$ of the substitutions presented must occur.

11. The problem is to prove that $x^4 + 1$ is irreducible. The equation

$$(x^2 + ax + b)(x^2 + cx + d) = x^4 + 1$$

can be solved for a, b, c, d to find $x^4 + 1 = (x^2 + x\sqrt{2} + 1)(x^2 - x\sqrt{2} + 1)$. Thus there is no factorization over $\mathbb{Q}$.

12. Roots of $x^4 + 1$ are roots of $x^8 - 1$. Therefore the analysis of §42 shows that the substitutions of the Galois group applied to these four roots can contain only those presented in §42, and no two substitutions of the group have the same effect on the roots of $x^4 + 1$. If the group had fewer than four elements then a root of $x^4 + 1$ would satisfy an equation of degree < 4 with rational coefficients.

13. Let S be the substitution which carries $abcd$ to $badc$, T the one which carries $abcd$ to $cdab$, and U the one which carries $abcd$ to $acdb$. Then $S^2 = T^2 = $ identity, $ST = TS$, and the subgroup is I, S, T, ST. Also $U^3 = I$. Computation gives $U^{-1}SU = ST$, $U^{-1}T U = S$. If these are written $SU = UST$, $TU = US$, they show that any $S^iT^jU = US^rT^s$ for suitable r, s. It follows that any composition of S, T, U can be reduced to the form $S^i T^j U^k$ for $i = 0$ or 1, $j = 0$ or 1, $k = 0$, 1, or 2. These are the twelve substitutions of the group. Since

$$(S^i T^j U^k)^{-1}(S^u T^v)(S^i T^j U^k) = U^{-k}T^{-j}S^{-i}S^u T^v S^i T^j U^k = U^{-(k-1)}U^{-1}S^u T^v U\, U^{k-1}$$

$$= U^{-(k-1)}S^r T^s U^{k-1},$$

it can be seen that every such substitution can be reduced to the form $S^a T^b$. Therefore the subgroup is normal by Exercise 3.

14. Let S be such that every element of G is a power of S. Let n be the least positive integer for which $S^n = I$. If k is a positive integer with $S^k = I$ and if d is the greatest common divisor of n and k then $d = An + Bk$, by the Euclidean algorithm, and $S^d = (S^n)^A (S^k)^B = I$. Thus $d \geq n$. Therefore $d = n$ and n divides k. It follows that $S^j = I$ if and only if $j \equiv 0$ mod n. The same argument shows that if H is a subgroup of G and if m is the least positive integer for which S^m is in H then S^j is in H if and only if $j \equiv 0$ mod m, i.e. if and only if S^j is a power of S^m.

15. The subgroup divides a presentation of G into two presentations of the subgroup H. If T is any element of G that is not in H, then T applied to an arrangement in either presentation of H must carry it to an arrangement in the other presentation since otherwise $T \in H$. The two presentations of course contain equal numbers of arrangements, so this shows that T carries either presentation to the other. Therefore H is normal.

16. Since $(b/a)^p = k/k = 1$, what is to be shown is that any root of $x^p = 1$ is a power of α. If $i > j$ and $\alpha^i = \alpha^j$ then $\alpha^{i-j} = 1$. If $i - j$ is not divisible by p then $Ap + B(i - j) = 1$ for some A and B and $\alpha = \alpha^{Ap + B(i-j)} = 1^B = 1$, contrary to assumption. Therefore $1, \alpha, \alpha^2, \ldots, \alpha^{p-1}$ are all distinct. Use induction and the Remainder Theorem to prove that a polynomial of degree n has at most n distinct roots.

17. $t = Aa + Bb + \ldots$ (for simplicity—the same proof applies to Galois resolvents t that are not linear). Thus t is a root of $X - A\phi_a(X) - B\phi_b(X) - \ldots = 0$. By Lemma I, so is t'. Thus $t' = A\phi_a(t') + B\phi_b(t') + \ldots$, as was to be shown.

18. Let elements of the Galois group be viewed as automorphisms of $K(a, b, c, \ldots) = K(t)$. Denote them by $S_1, S_2, \ldots, S_m$. Then $G(X) = \prod(X - S_i t)$ has coefficients invariant under the Galois group and therefore in K. $G(X)$ is irreducible because any factor must be divisible by at least one $X - S_i t$ and therefore by all. (If $S_i t = S_j t$ then $i = j$ because if $S_k t = t$ then $S_k = $ identity.) The expressions $a = \phi_a(t)$, $b = \phi_b(t), \ldots$ exist by virtue of the definition of "primitive element." Application of $S_1, S_2, \ldots, S_m$ to these expressions gives the presentation of the Galois group.

Sixth Exercise Set

1. The arrangements abc, bca, cab, present a normal subgroup of index 2 in which the identity is a normal subgroup of index 3. Let the three subgroups of this sequence be written $G_6 \supset G_3 \supset G_1$, where the subscripts denote the number of elements. To implement Galois' method, one must choose, for each consecutive pair of groups in the sequence—in this case two pairs—a polynomial in the roots that is invariant under the smaller group but not under the larger. In this case, one can choose $a^2 b + b^2 c + c^2 a$ and a. The general method then calls for setting $\Delta = a^2 b + b^2 c + c^2 a - ab^2 - bc^2 - ca^2$. Then Δ^2 is invariant under G_6 and can therefore be expressed in terms of the coefficients of the equation. Moreover, any polynomial in the roots $F(a, b, c)$ which is invariant under G_3 can be written $r + s\Delta$ where r and s are invariant under G_6. Finally the general method calls for setting $U = a + \alpha b + \alpha^2 c$. Then U^3 is invariant under G_3 and every polynomial in the roots can be written $x + yU + zU^2$ where x, y, and z are polynomials in the roots that are invariant under G_3. Then Δ and U are radicals (Δ^2 is known and $U^3 = r + s\Delta$ where r, s are known) and every polynomial in the roots can be written in the form $r_1 + s_1\Delta + r_2U + s_2\Delta U + r_3U^2 + s_3\Delta U^2$ where the r's and s's are known. In particular, a, b, and c themselves can be written in this form.

In the explicit solution, assume for the sake of simplicity that $a + b + c = 0$, i.e. the given cubic has the form $x^3 + px + q = 0$. Then $\Delta^2 = -4p^3 - 27q^2$ can be found by a relatively easy computation in symmetric functions. (Exercise 25 of the First Set.) Another computation gives

$$U^3 = a^3 + b^3 + c^3 + 6abc + 3\alpha(a^2b + b^2c + c^2a) + 3\alpha^2(ab^2 + bc^2 + ca^2).$$

To write this in the form $U^3 = r + s\Delta$, interchange a and b and add to find

$$2r + s(\Delta - \Delta) = 2a^3 + 2b^3 + 2c^3 + 12abc$$
$$+ (3\alpha + 3\alpha^2)(a^2b + b^2c + c^2a + ab^2 + bc^2 + ca^2),$$

$r = a^3 + b^3 + c^3 + 6abc - \frac{3}{2}$ every $a^2b = -3q - 6q - \frac{3}{2} \cdot 3q = -27q/2$. Similarly, $2s\Delta = 3\alpha\Delta - 3\alpha^2\Delta$, $s = 3(\alpha - \alpha^2)/2 = 3\sqrt{-3}/2$. Thus U and Δ can be explicitly expressed as radicals

$$\Delta = \sqrt{-4p^3 - 27q^2}.$$

$$U = \sqrt[3]{(-27q + 3\Delta\sqrt{-3})/2}.$$

Since

$$3a = (a + b + c) + U + (a + \alpha^2 b + \alpha c) = U + U',$$

$3b = \alpha^2 U + \alpha U'$, $3c = \alpha U + \alpha^2 U'$, to express a, b, c (and hence all polynomials in a, b, c) in terms of Δ, U, and known quantities, it will suffice to so express U'. Now $UU' = a^2 + b^2 + c^2 - (ab + bc + ca) = -3p$ is known. Thus $U' = -3p/U$. To avoid division by U, multiply numerator and denominator by U^2 to find $U' = -3pU^2/(-27q + 3\Delta\sqrt{-3})2^{-1}$. Now multiply numerator and denominator by $-27q - 3\Delta\sqrt{-3}$ and simplify to find

$$U' = -U^2(9q + \Delta\sqrt{-3})/6p^2.$$

(The case $p = 0$ does not occur, of course, in the *general* cubic, i.e. when the coefficients are letters. If $p = 0$ then the equation has the simple solution $\alpha^i\sqrt[3]{q}$, where $i = 0, 1, 2$.) Since $\sqrt{-3} = \alpha - \alpha^2$, the solution can then be written

$$a = \frac{1}{18p^2}[6p^2U - 9qU^2 - (\alpha - \alpha^2)\Delta U^2],$$

$$b = \frac{1}{18p^2}[\alpha^2 6p^2U - 9q\alpha U^2 - \alpha(\alpha - \alpha^2)\Delta U^2],$$

$$c = \frac{1}{18p^2}[\alpha 6p^2U - 9q\alpha^2 U^2 - \alpha^2(\alpha - \alpha^2)\Delta U^2].$$

If the equation is of the more general form $x^3 + rx^2 + sx + t$, then the solution can be derived from the special case by setting $p = s - (r^2/3)$, $q = t - (rs/3) + (2r^3/27)$ in the above formulas and subtracting $r/3$ from each of the roots.

2. Galois' presentation (3) of §40 gives $G_{24} \supset G_{12} \supset G_4 \supset G_2 \supset G_1$. Since a subgroup of index 2 is normal (Exercise 15 of the Fifth Set) one needs only to show that G_4 is normal in G_{12}. This is Exercise 13 of the Fifth Set. Again, one needs to choose for each successive pair of subgroups in the sequence a polynomial in the roots invariant under the smaller but not under the larger. Such polynomials are, for example, $\Delta =$ Vandermonde determinant $= \pm(a - b)(a - c)(a - d)(b - c)(b - d)(c - d)$, $ab + cd$, $a + b$, a. Again, the general method calls for setting $U = (ab + cd) + \alpha(ac + db) + \alpha^2(ad + bc)$, $V = a + b - c - d$, $W = a - b$, and writing an arbitrary polynomial in a, b, c, d as a polynomial in Δ, U, V, W with known coefficients and in expressing Δ, U, V, W as radicals.

Δ^2 is symmetric in a, b, c, d and therefore can be expressed in terms of the coefficients of the given equation. For the explicit expression, see the next exercise. Now $U = P + \alpha Q + \alpha^2 R$ where $P = ab + cd, Q = ac + bd, R = ad + bc$. The preceding exercise then gives explicit expressions for P, Q, R in terms of the elementary symmetric functions in P, Q, R, two explicit radicals, U and Δ, and the cube root of unity α. Since the elementary symmetric functions in P, Q, R are symmetric in a, b, c, d (this is a special feature of the choice $P = ab + cd$ and would not be the case if the equally valid choice $P = (a - b)(c - d)$ had been made) this expresses P, Q, R in terms of radicals U, Δ and the coefficients of the given equation. Explicitly, $P + Q + R = \sigma_2$. Moreover, $PQ = a^2bc + b^2ad + c^2ad + d^2bc$, from which it is clear that $PQ + QR + RP = \sum a^2bc$. Now $\sigma_1\sigma_3 = \sum a^2bc + 4abcd$, so $PQ + QR + RP = \sigma_1\sigma_3 - 4\sigma_4$. $PQR = \sum a^3bcd + \sum a^2b^2c^2$. $\sum a^3bcd = \sigma_4 \sum a^2 = \sigma_4(\sigma_1^2 - 2\sigma_2)$ and $\sigma_3^2 = \sum a^2b^2c^2 + 2 \sum a^2b^2cd = \sum a^2b^2c^2 + 2\sigma_4\sigma_2$, so $PQR = \sigma_1^2\sigma_4 + \sigma_3^2 - 4\sigma_2\sigma_4$. (The Δ in the formulas of the preceding exercise differs from the Δ above in its definition; the fact that they are equal is the basis of the next exercise.)

Next,

$$V^2 = a^2 + 2ab + b^2 + c^2 + 2cd + d^2 - 2(a + b)(c + d)$$
$$= (a^2 + b^2 + c^2 + d^2) + 4(ab + cd) - 2\sum ab = 4P + \sigma_1^2 - 4\sigma_2.$$

Thus V is the square root of an expression in Δ and U. The expression of W as the square root of an expression in $\Delta, U,$ and V can be obtained by introducing $V_2 = a - b - c + d$ and $V_3 = a - b + c - d$; then $V_2^2 = 4R + \sigma_1^2 - 4\sigma_2$, $V_3^2 = 4Q + \sigma_1^2 - 4\sigma_2$, and $VV_2V_3 = \sum a^3 + 2\sigma_3 - \sum a^2b$. From $\sigma_1^3 = \sum a^3 + 3\sum a^2b + 6\sigma_3$ and $\sigma_1\sigma_2 = \sum a^2b + 3\sum abc$, it follows that $VV_2V_3 = \sigma_1^3 - 4\sigma_1\sigma_2 + 8\sigma_3$. Then $2W = V_2 + V_3$ and $4W^2 = V_2^2 + 2(VV_2V_3)V^{-1} + V_3^2$ is an expression in $U, \Delta,$ and V. Of course division by V can be avoided by writing

$$V^{-1} = VV^{-2} = V(4P + \sigma_1^2 - 4\sigma_2)^{-1} = V(4Q + \sigma_1^2 - 4\sigma_2)(4R + \sigma_1^2 - 4\sigma_2)/D$$

where the denominator D is $(4P + X)(4Q + X)(4R + X) = X^3 + 4X^2(P + Q + R) + 16X(PQ + QR + RP) + 64PQR$, where $X = \sigma_1^2 - 4\sigma_2$. This gives W as an explicit radical. Finally, $4a = (a + b + c + d) + V + 2W$ gives a in terms of radicals. Then $b = a - W$ expresses b. The other two roots can be expressed in terms of $W' = c - d$, and this can be expressed in terms of the radicals Δ, U, V, W by means of $W' = (Q - R)/W$ where, as before, division by W can be expressed in terms of multiplication by expressions in Δ, U, V, W and division by a polynomial in $\sigma_1, \sigma_2, \sigma_3, \sigma_4$.

3. $P - Q = (a - d)(b - c)$. Thus

$$(P - Q)(Q - R)(R - P) = \pm(a - b)(a - c)(a - d)(b - c)(b - d)(c - d) =$$
$$\pm \text{ Vandermonde determinant of } a, b, c, d.$$

The discriminant of the quartic is therefore equal to the discriminant of the cubic

$$X^3 - (P + Q + R)X^2 + (PQ + QR + RP)X - PQR = 0.$$

By the formulas of the preceding exercise, this is the cubic $x^3 - \sigma_2 x^2 + (\sigma_1\sigma_3 - 4\sigma_4)x - (\sigma_1^2\sigma_4 + \sigma_3^2 - 4\sigma_2\sigma_4)$. If the quartic has the form $x^4 + \alpha x^2 + \beta x + \gamma$ where $\sigma_1 = 0$ then, by the formula of §13, this discriminant is

$$-128\alpha^2\gamma^2 + 16\alpha^4\gamma - 4\alpha^3\beta^2 + 144\,\alpha\beta^2\gamma + 256\gamma^3 - 27\beta^4.$$

4. The suggested lemma proves the theorem, because a succession of extensions like the one of the proposition of §44 gives, by assumption, a root of the given irreducible polynomial and hence an irreducible (linear) factor whose roots can be expressed in terms of radicals. Thus all roots of one of the irreducible factors at the next-to-last stage can be expressed in terms of radicals, and so forth, until one arrives at the desired conclusion. The lemma says, in effect, that if all roots of g_1 can be expressed in terms of radicals then so can all roots of g_i for all i. In other words, if the Galois group of g_1 over K' is solvable then so is the group of g_i over K'. If $g_1 \neq f$ then adaptations of arguments of §44 show that $f(x) = g(x, r)g(x, \alpha r)g(x, \alpha^2 r) \ldots g(x, \alpha^{p-1} r)$ where $g(x, y)$ is a polynomial in two variables with coefficients in K and where the factors on the right are irreducible over $K' = K(r)$. Moreover, $g(x, \alpha r) = g(x, Tr)$ for some substitution T of the Galois group of $f(x) = 0$ over K because f is irreducible over K. Thus, given any two of the polynomials $g(x, \alpha^j r)$ there is a suitable power T^j of T which carries roots of one to roots of the other and leaves elements of K fixed. If u is a Galois resolvent of the first (over K') then $T^j u$ is a Galois resolvent of the second. When u is used to construct a presentation of the Galois group of the first, T^j applied to this construction gives a presentation of the Galois group of the second. Therefore a sequence of subgroups of the required type for the first Galois group implies one for the second.

5. The Galois group of $(x^p - 1)/(x - 1)$ over any field (of characteristic 0) is a subgroup of the cyclic group of order $p - 1$ (see §42). Therefore the Galois group is solvable. If K contains p_ith roots of unity for all the indices that occur in this sequence of subgroups — which is guaranteed if it contains p_ith roots of unity for all prime factors p_i of $p - 1$ — then, by the proposition of §46, all pth roots of unity can be obtained by adjoining radicals to K. Thus it suffices to show that all p_ith roots of unity can be obtained by radical adjunctions to $\mathbb{Q}$, where $p_i | p - 1$. An induction now sets in. See §65.

6. As in the proof of §44, $G(X) = H(X, r^{(i)})Q(X, r^{(i)})$ for all m roots $r, r', r'', \ldots, r^{(m-1)}$ of the auxiliary equation. Then

$$G(X)^m = h(X)q(X)$$

where

$$h(X) = H(X, r)H(X, r')H(X, r'')\ldots.$$

This gives $h(X) = \text{const.}\ G(X)^j$ for some j, from which $m = j \cdot (\deg G/\deg H)$, as desired.

7. To adjoin all roots is the same as to adjoin a Galois resolvent u of $g(x) = 0$ over K. Any other root u' of the irreducible polynomial satisfied by u is a rational function of u. Therefore the polynomials $H(X, u)$, $H(X, u')$, $H(X, u'')$, ... all have coefficients in the same field $K(u)$ and are irreducible over that field. Therefore, any two of them which have a root in common have identical roots. As in §44, their product is $\text{const.} \cdot G(X)^j$ for some j. Since the H's have simple roots, every root of G is a root of exactly j of the H's. Therefore the u's can be rearranged, if necessary, so that

$$G(X) = \text{const. } H(X, u)H(X, u')H(X, u'') \cdots H(X, u^{(\mu/j)}),$$

where μ is the number of u's. This partitions the presentation of the old group into μ/j presentations of the new, and the proof that the new is a normal subgroup of the old is as in §44.

8. See §54 on unique factorization of polynomials and §60 on norms of polynomials.

9. Any substitution of the group over $K(u)$ is obviously in the group over K, and it leaves u fixed by Proposition I. Conversely, if S is a substitution in the group over K, if $Su = u$, and if $H(X)$ is a polynomial in X with coefficients in $K(u)$ of which t is a root then application of S to $H(t) = 0$ gives $H(St) = 0$. Thus the group over $K(u)$ contains the substitution which carries $\phi_a(t), \phi_b(t), \phi_c(t), \ldots$ to $\phi_a(St), \phi_b(St), \phi_c(St), \ldots$ and this is the substitution S.

10. By Exercise 4 of the Fifth Set, the Galois group of $fg = 0$ over $K(b_1)$ acts transitively on the roots of f_j. By Exercise 9, this Galois group contains the substitutions of the Galois group of $fg = 0$ over K that leave b_1 fixed. Conversely, any such substitution carries roots of f_k to roots of f_k. The suggested rule assigns at least one index $k = 1, 2, \ldots, \mu$ to each factor of g over $K(a_1)$, and assigns each index to at least one factor. It needs to be shown that it never assigns different indices to the same factor. By the preceding, this amounts to showing that if S_1 and S_2 are given then there is a T_1 with $T_1 b_1 = b_1$ and $T_1 S_1 a_1 = S_2 a_1$ if and only if there is a T_2 with $T_2 a_1 = a_1$ and $T_2 S_1^{-1} b_1 = S_2^{-1} b_1$. This follows from the formula $T_2 = S_2^{-1} T_1 S_1$. Since the S's act transitively on the a's (Exercise 4 of the Fifth Set) the number of S's that carry a_1 to a_j is the same for all j, call it N. Then $N \cdot \deg f = $ number of S's and $N \cdot \deg f_k = $ number of S's that carry a_1 to a root of f_k. But Sa_1 is a root of f_k if and only if $S^{-1} b_1$ is a root of g_k. Thus $\deg f_k / \deg f = $ proportion of S's that carry a_1 to a root of $f_k = $ proportion of S's such that $S^{-1} b_1$ is a root of $g_k = \deg g_k / \deg g$.

11. K' can be obtained from K by adjoining a Galois resolvent u of g over K. Let $G(X) = 0$ be the irreducible polynomial with coefficients in K of which u is a root. If K'' is an extension of K then the irreducible factors of G over K'' all have the same degree, namely, the number of substitutions in the Galois group of $g = 0$ over K''. Apply Dedekind's theorem in the case $K' = K(u)$, $K'' = K(a)$ where a is a root of f.

12. The solution of §5 is $A - B$ where A and B are determined by $27A^6 + 27qA^3 - p^3 = 0, 3AB = p$. The solution of Exercise 1 is $(U + U')/3$ where U is an explicit element of a field obtained by adjoining first a square root, then $\sqrt{-3}$, and then a cube root, and where $UU' = -3p$. In the new field, the quadratic equation satisfied by A^3 has two solutions, one of which is $(U/3)^3$. If $A = U/3$ then $B = -U'/3$ and the solutions coincide. If $A = \alpha^j U/3$, then the solution coincides with one of the other solutions in Exercise 1. The product of the two solutions for A^3 is $(-p/3)^3 = (UU'/9)^3$, so the other solution is $(U'/3)^3$ and the three solutions are again given by Exercise 1.

13. An element of K' in $K(a, b, c, \ldots)$ but not in K would be moved by some element of the Galois group over K (Proposition I) but by no element of the Galois group over K', so the two groups would not coincide. Conversely, if the two groups do not coincide, the argument of §46 shows that some element of $K(a, b, c, \ldots)$ not in K is invariant under the smaller group, i.e. is in K'.

Seventh Exercise Set

1. If $\mu > 0$, then, since f_1 divides $f_1 f_2 \ldots f_\mu = g_1 g_2 \cdots g_\nu$, f_1 divides either g_1 or $g_2 \ldots g_\nu$. If it divides $g_2 \ldots g_\nu$ it divides either g_2 or $g_3 g_4 \ldots g_\nu$, etc. Thus f_1 divides g_1 or g_2 or $\ldots$ or g_ν. In particular, $\nu > 0$. Renumber the g's so that f_1 divides g_1, say $g_1 = q_1 f_1$. Repetition of this argument μ times gives $\nu \geq \mu$, $g_i = q_i f_i$ ($i = 1, 2, \ldots, \mu$) for some rearrangement of the g's, and consequently $1 = q_1 q_2 \ldots q_\mu g_{\mu+1} \cdots g_\nu$. Since an irreducible polynomial can not divide 1, $\nu > \mu$ is impossible. The q's are units because they divide 1.

2. The proof of Corollary 1 given in the proof of the theorem deduces it from Gauss's lemma, that is, from the case deg $F = 0$ of the theorem.

3. Let $F(a, x) \in K[a, x]$ be irreducible. First suppose F has degree 0 in x, i.e. $F \in K[a]$. Then, directly from the definitions, F divides $G = G_0 + G_1 x + G_2 x^2 + \cdots + G_n x^n \in K[a, x]$, where $G_i \in K[a]$, if and only if it divides each G_i as an element of $K[a]$. Since F is irreducible in $K[a]$, it is prime (§54). The argument of §57 then shows that F divides GH only if it divides G or H. Now let F contain x. If it were reducible as a polynomial in x with coefficients in $K(a)$ (K with a transcendental element a adjoined), say $F = gh$, then multiplication by a suitable element of $K[a]$ (polynomials in a with coefficients in K) would give $f(a)F(a, x) = G(a, x)H(a, x)$, where f, G, H are polynomials in the indicated variables with coefficients in K. Factor f into irreducibles and divide by them one-by-one to find $F(a, x) = G_1(a, x)H_1(a, x)$ where G_1, H_1 are polynomials in a and x, contrary to assumption. Thus F is irreducible over $K(a)$. Irreducible polynomials are prime, so if F divides $GH \in K[a, x]$ then F divides G or H as a polynomial in x with coefficients in $K(a)$, say $G = FQ$ where Q has coefficients in $K(a)$. Then the denominators of Q can be cleared to give $g(a)G(a, x) = F(a, x)Q_1(a, x)$ where g and Q_1 are polynomials in the indicated variables. Division by the irreducible factors of g gives $G(a, x) = F(a, x)Q_2(a, x)$, i.e. F divides G in $K[a, x]$. Corollary 1 was proved above. Corollary 2 follows as in Exercise 1.

4. If the given polynomial is $f(x) = A_n x^n + \cdots + A_0$ where the A's are rational functions of a, and if $D(a)$ is the product of the denominators of the A's then $D(a) \cdot f(x)$ is a polynomial in two variables, say $F(a, x)$, and $f(x) = F(a, x)/D(a)$ is a representation in the desired form. The second statement follows from the observation that a unit times an irreducible polynomial is irreducible.

5. $\bar{g}$ is a sum of terms ct^j where $c \in K$, $j \geq 0$. Given such a term, write $j = rN + s$ where $0 \leq s < N$. The corresponding term of g is $ca^r x^s$.

6. The values at $x = -2, -1, 0, 1, 2$ are $-3, -2, -3, 18 = 2 \cdot 3^2, 133 = 7 \cdot 19$. The leading coefficient, which can be regarded as the value at ∞, is 2, so any factor must have leading coefficient $\pm 1, \pm 2$. The values at $-3, -4, \ldots$ are easily seen to be positive, so there is no integer root, and therefore no factor $x - a$. The only possible factors $2x - a$ would correspond to roots at halfintegers, and these could occur only in the intervals $[-3, -2]$, $[0, 1]$ where the sign changes. However, neither $-2\frac{1}{2}$ nor $\frac{1}{2}$ is a root. Thus the polynomial is irreducible unless it has a factor of the form $2x^2 + qx + p$. Here $p = -3, -1, 1$, or 3. Try $p = 1$. If v_1 and v_2 are the values of this polynomial at -1 and -2 respectively then $q = 3 - v_1$ and $q = (9 - v_2)/2$. Since $v_1 | 2$ and $v_2 | 3$, the first gives $q = 5, 4, 2, 1$ and the second gives $q = 6, 5, 4, 3$. The value at 1 of $2x^2 + qx + 1$ is $3 + q$ does not divide 18 for $q = 4$ or 5, so there is no factor $2x^2 + qx + 1$. The polynomial $2x^2 + 2x - 1$ has acceptable values at $-2, -1, 0, 1$, but not at 2. The desired factorization is $(2x^2 + 4x + 3)(x^2 + 2x - 1)$.

7. The determinant of the matrix whose rows are 1 2 0, 0 1 2, and $-4 -2$ 1, in that order, is -11. $(1 + 2a)(1 + 2b)(1 + 2c) = 1 + 2(a + b + c) + 4(ab + bc + ca) + 8abc = 1 + 2\cdot 0 + 4\cdot 1 + 8(-2) = -11$.

8. Let d be the least common denominator of the coefficients of g so that $dg = G$ where G has integer coefficients and no prime divisor of d divides G. Similarly, let $eh = H$. Since $gh = F$ has integer coefficients, de divides $deF = GH$. A prime factor of d cannot divide G and must therefore divide H. In this way, all prime factors of d can be divided one-by-one into H, and it follows that all coefficients b_j of H are divisible by d. Similarly, all coefficients a_i of G are divisible by e. Thus, a coefficient a_i/d of g times a coefficient b_j/e of h gives a product $(a_i/e)(b_j/d)$, which is an integer.

9. g is invariant under the Galois group and therefore has coefficients in K. The norm can be defined using any basis of $K(a)$ over K because $\det(Q^{-1}MQ) = \det(M)$. In particular, the basis $1, b, b^2, \ldots, b^{n-1}$ can be used, and application to $f \cdot a^i = \sum c_{ij}a^j$ of an element of the Galois group which carries a to b gives $f_b \cdot b^i = \sum c_{ij}b^j$, from which $Nf = \det(c_{ij}) = Nf_b$. Thus $Ng = Nf \cdot Nf_b \cdot Nf_c \cdot \ldots = (Nf)^n$. The key equation $N_t f = (N_a f)^k$ follows from the fact that the matrix whose determinant is $N_t f$ is made up of k copies down the main diagonal of the matrix whose determinant defines $N_a f$, one copy corresponding to $t^j, at^j, a^2 t^j, \ldots, a^{n-1}t^j$ for each $j = 0, 1, \ldots, k - 1$.

10. It is to be shown that four conditions are equivalent: (A) ϕ divides ϕ'. (B) $\phi' = 0$. (C) $\phi(x) = \psi(x^p)$. (D) In a splitting field for ϕ, two roots of ϕ coincide. (A) $\Rightarrow$ (B) because a polynomial divides no polynomial of lower degree except 0. (B) $\Rightarrow$ (C) because the formula for the derivative shows that $\phi' = 0$ only when powers of x not divisible by p all have coefficient 0. (C) $\Rightarrow$ (D) because if a_1 is a root of ϕ then a_1^p is a root of ψ, $x - a_1^p$ divides $\psi(x)$, and $x^p - a_1^p = (x - a_1)^p$ divides $\phi(x)$, so a_1 is a p-fold root, at least, of ϕ. (D) $\Rightarrow$ (A) because if $(x - a_1)^2$ divides ϕ then $x - a_1$ divides $d = \text{g.c.d.}$ (ϕ, ϕ'); because ϕ is irreducible, $d = c$ or $c\phi$ where $c \in K$ is nonzero; because $x - a_1$ divides d, $d \neq c$, $d = c\phi$, and ϕ divides ϕ'.

11. For any $n \in F_p$, Fermat's theorem gives $n^p = n$, that is, every element of F_p is *its own* pth root. Suppose a is the root of an irreducible polynomial of degree 2. Then the elements of $K(a)$ can be written uniquely in the form $u + va + wa^2$ where $u, v, w \in K$. It is to be shown that for every $u, v, w \in K$ there are $x, y, z \in K$ such that $x^p + y^p a^p + z^p a^{2p} = u + va + wa^2$. This is the same, since every element of K can be written as a pth power, as showing that there are $r, s, t \in K$ with $r + sa^p + ta^{2p} = u + va + wa^2$, that is, 1, a^p, a^{2p} are a basis of $K(a)$. This is true unless there are $r, s, t \in K$, not all zero, with $r + sa^p + ta^{2p} = 0$. Such a triad would give $(x + ya + za^2)^p = 0$, $x + ya + za^2 = 0$ with x, y, z not all zero, which is impossible. The same argument applies when a is the root of an irreducible polynomial of any degree. To prove (3) note that if $\phi(x) = \psi(x^p)$, say $\phi(x) = x^{kp} + \cdots + a_j x^{jp} + \cdots + a_k$, then $\phi(x) = \omega(x)^p$ where $\omega(x) = x^k + \cdots + b_j x^j + \cdots + b_k$ and $b_j^p = a_j$. Therefore $\phi(x) = \psi(x^p)$ implies ϕ is *not* irreducible. (4) follows from the observation that if a is transcendental over K then $a = f(a)^p/g(a)^p$ is impossible.

12. $N(F)^p = N(F^p)$ (norm of a product) $= (F^p)^p$ (the coefficients of F^p are in K). The argument of Exercise 9 then gives $N(F) = F^p$.

13. Existence is proved by the construction indicated in §53. By (1), the field constructed in this way can be mapped onto any $K(t')$. By (2), this map is an isomorphism.

14. M_{X-F} is a matrix whose entries are polynomials with coefficients in K. Let v be the column vector $1, a, a^2, \ldots, a^{n-1}$. Then $M_{X-F} \cdot v = (X - F) \cdot v$ where the multiplication on the right signifies multiplication of each entry of v by $X - F$. Substitution of F for X gives $\hat{M} \cdot v = 0$, from which $\det(\hat{M}) = 0$, as was to be shown.

Eighth Exercise Set

1. Suppose $v_1, v_2, \ldots, v_n$ span V. If they are not linearly independent then one of them, say v_n, can be expressed as a linear combination of the others. Thus $v_1, v_2, \ldots, v_{n-1}$ span. Repeat until the spanning vectors are linearly independent. (From a constructive point of view, one must make some mild assumption about the way V is presented in order to know that, given a finite set of vectors, one can either find a nontrivial linear relation among them or show there is none.) A basis of n elements amounts to an isomorphism of V with the vector space K^n of n-tuples of elements of K. A mapping of K^n to K^m which respects the vector space structures is represented by an $m \times n$ matrix. It is to be shown that such a mapping can be invertible only if $m = n$. Prove that m homogeneous linear equations in n unknowns have a nontrivial solution when $m < n$ (solve one equation for one unknown in terms of the others and substitute to reduce to $m - 1$ equations in $n - 1$ unknowns) so that $K^n \to K^m$ cannot be one-to-one if $m < n$.

2. Any element x of L can be written in the form $x = \sum a_i u_i$ where $a_i \in K'$. Each a_i can be written $a_i = \sum b_{ij} s_j$ where $b_{ij} \in K$. Thus $x = \sum b_{ij} s_j u_i$. If $x = 0$ then necessarily all a_i are 0, from which all b_{ij} must be 0.

3. Multiplication by $j \mod p$ is one-to-one (if $j\mu \equiv j\nu \mod p$ then p divides $j(\mu - \nu)$ and $\mu \equiv \nu \mod p$) and therefore onto. Thus for each $i = 1, 2, \ldots, p - 1$ there is one and only one $k = 1, 2, \ldots, p - 1$ such that $jk \equiv i \mod p$. Since $g(a^{jk}) = g(a^i)$, this shows that $G(a^j) = G(a)$.

4. If the equation is solvable by radicals, the elements of the Galois group can be represented as transformations of the form $i \mapsto ri + s$ ($r \not\equiv 0 \mod p$). If i and j are both fixed under such a transformation and $i \not\equiv j \mod p$ then $r \equiv 1, s \equiv 0 \mod p$. Thus, after two distinct roots of the equation are adjoined, the Galois group is reduced to the identity element, that is, all roots are rational functions of the two that were adjoined. Conversely, suppose the Galois group has the property that each of its elements, except for the identity, leaves at most one root fixed. Using the fact that the group acts transitively on a set with p elements, one can show that the number of elements is divisible by p. Therefore the group contains an element of order p, say S. The proof of the theorem shows that it will suffice to show that the powers of S are a normal subgroup of the Galois group. To this end, let T be an element of the Galois group. Then TST^{-1} is an element of order p. Let $a_1, a_2, \ldots, a_p$ be the roots of f and for each $i = 1, 2, \ldots, p$ let $\mu(i)$ be the integer mod p defined by $TST^{-1}a_i = S^{\mu(i)}a_i$. Since $\mu(i) \not\equiv 0 \mod p$, the p values i correspond to $p - 1$ values $\mu(i)$ and there are distinct values of i with the same $\mu(i)$, that is, distinct roots a_i that are left fixed by $S^{-\mu}TST^{-1}$ for some μ. By assumption, then, $S^{-\mu}TST^{-1}$ is the identity, that is $TST^{-1} = S^\mu$, and the powers of S are a normal subgroup, as desired.

5. Let f be the given polynomial with coefficients in K and let f_1 be an irreducible factor of f over $K' = K(r)$. As in §44, $f_1(x, r)f_1(x, \alpha r) \ldots f_1(x, \alpha^{p-1}r)$ is a constant times a power of $f(x)$, say $f(x)^j$. Two of the factors $f_1(x, \alpha^i r)$ have a root in common if and only if they have the same roots. Since f and f_1 have simple roots, each root of f is a root of exactly j of the factors $f_1(x, \alpha^i r)$. This partitions the p factors into subsets each containing j factors. If $j = 1$ then f is irreducible over K'. If $j = p$ then f has been factored in the required way.

6. Van der Waerden gives the following proof. Over the splitting field, $f(x) = \prod_i (x - \alpha^i r)$ where $r^p = k$ and α is a primitive pth root of unity. If $g(x)$ is a nontrivial factor of $f(x)$ with leading coefficient one and if b is its constant term then $b = \alpha^v r^\mu$ where μ and v are positive integers and $\mu < p$. Then $b^p = r^{\mu p} = k^\mu$. There are integers A and B with $Ap + B\mu = 1$. Then $k = k^{Ap} k^{B\mu} = (k^A b^B)^p$, q.e.d. To prove the theorem more straightforwardly, note that it amounts to saying that a subgroup of the $p(p - 1)$-element group of linear transformations $i \mapsto ri + s$ ($r \not\equiv 0 \bmod p$) of the integers mod p is transitive unless some integer mod p is invariant under all transformations in the subgroup. Let H be a subgroup of this group, let j be an integer, and let $\{j_1, j_2, \ldots, j_\mu\}$ be the orbit of j under H, that is, all classes of integers mod p which contain integers $S(j)$ where $S \in H$. If $S: i \mapsto ai + b$ is in H, then, mod p, the integers $aj_k + b$, are a rearrangement of the j's. Therefore, summation gives $a\sigma + \mu b \equiv \sigma \bmod p$, where $\sigma = j_1 + j_2 + \cdots + j_\mu$. If $\mu < p$, one can divide this congruence by μ mod p—which is to say, multiply by the B above—to find that the class of $B\sigma$ is fixed under S for all $S \in H$.

7. Let K be the given coefficient field and let $L = K(x_1, x_2, \ldots, x_n)$ be the field obtained by the (transcendental) adjunction of n variables $x_1, x_2, \ldots, x_n$ to K. Elements of L can be expressed as quotients of polynomials. Because numerator and denominator can be multiplied by the conjugates of the denominator (the polynomials obtained by permuting the variables of the denominator) one can assume without loss of generality that the denominator is symmetric. Thus elements of L can be written in the form $\sum R(\sigma) x_1^{c_1} x_2^{c_2} \ldots x_n^{c_n}$ where the $R(\sigma)$ are *rational* functions in $\sigma_1, \sigma_2, \ldots, \sigma_n \in L$, and this representation is unique. The rational functions $R(\sigma)$ therefore form a subfield K'' of L over which L is a vector space of dimension $n!$. Clearly L is the splitting field of the polynomial $(X - x_1)(X - x_2) \ldots (X - x_n) = X^n - \sigma_1 X + \sigma_2 X^2 - \cdots \pm \sigma_n$ with coefficients in K''. Since the degree of the splitting field is the order of the Galois group, the Galois group contains all $n!$ substitutions of the roots $x_1, x_2, \ldots, x_n$.

8. If f is reducible over $\mathbb{Q}$ it is reducible over $\mathbb{Z}$. Since $f(x) \equiv x^n \bmod p$, the only factors mod p are nonzero constants times powers of x mod p. If f has a nontrivial factorization over $\mathbb{Z}$ the constant terms of both factors must therefore both be 0 mod p. This implies that the constant term of f is divisible by p^2, contrary to assumption. The coefficients of $f(x) = [(x + 1)^p - 1]/x$ satisfy the conditions so f is irreducible. Therefore so is $f(X - 1)$, because $f(X - 1) = g(X)h(X)$ implies $f(X) = g(X + 1)h(X + 1)$.

9. In the expansion of $(y + z)^p$ all binomial coefficients except the first and the last are divisible by p. Thus if the statement to be proved is true of F and G it is true of $F + G$. It is true of any monomial with coefficient ± 1. Every polynomial can be written as a sum of such monomials.

10. Let K' be an extension of K and let both be finite. What is to be shown is that $K' = K(t)$ for some t. Since every field with characteristic p contains the field F_p of integers modulo p, it will suffice to show that every finite field K' is of the form $F_p(t)$ for some t. Let q be the number of elements in K'. Then every nonzero element a of K' satisfies $a^{q-1} = 1$ and every element of K' is a root of $x^q - x = 0$. Now q is a power of p, namely, $q = p^n$ where $n = [K' : F_p]$. The derivative of $x^q - x$ is therefore -1, which implies that $x^q - x$ has no multiple roots. Thus $x^q - x = \prod (x - t_i)$ where t_i ranges over all q elements of K'. Every nonzero element t of K' is a $(q - 1)$st root of unity and $K' = F_p(t)$ if and only if t is a *primitive* $(q - 1)$st root of unity. For each divisor a of $q - 1$ let $\phi(a)$ denote the number of primitive ath roots of unity in K. Then $a = \sum \phi(j)$ where j runs over all divisors of a. Since these are the same equations satisfied by the Euler ϕ-function ($\phi(a)$ = number of positive integers less than a relatively prime to a) ϕ coincides with the Euler ϕ-function, and, in particular, $\phi(q - 1) > 0$.

11. Let $Q(x)$ be the greatest common divisor of f and f'. Let g be an irreducible factor of f over K, say $f = g^n h$ where $g \nmid h$. Then f' is divisible exactly by $n - 1$ times by g unless $ng' = 0$, i.e. unless $n = pv$ ($v > 0$) or $g' = 0$. If $g' = 0$ then g is a polynomial in x^p, and if $n = pv$ then $g(x)^n = g(x^p)^v$, so in both cases f is divisible by a polynomial in x^p. Otherwise the roots of g are simple roots of f/Q.

12. The argument of Exercise 11 gives $f(x) = Q(x)R(x^p)$, where Q has distinct roots which are not roots of $R(x^p)$. By Fermat's theorem $R(x^p) = S(x)^p$ where the coefficients of S are the pth roots of the corresponding coefficients of R. Thus every root of $f(x)$ is a root of $Q(x)S(x)$. If $S(x)$ does not have distinct roots, repeat. The pth power map $x \mapsto x^p$ is a homomorphism of K to itself (Fermat's theorem) that is one-to-one ($x^p = 0$ implies $x = 0$). If K is finite it must be onto.

13. If b is a root of $x^p - x - a$ so is $b + 1$. This gives p distinct roots. The arguments of §44 are easily modified to prove the proposition.

14. What is needed is an analog of the "Lagrange resolvent" of §46. If $\theta_1, \theta_2, \ldots, \theta_p = \theta_0$ are determined, as in §46, with $S\theta_i = \theta_{i+1}$ then the sum $s = \theta_1 + 2\theta_2 + \cdots + (p - 1)\theta_{p-1}$ satisfies $Ss = s - (\theta_1 + \theta_2 + \cdots + \theta_p) = s - k$ where k is in K. With $\psi = k^{-1}s$ this gives $S\psi = \psi - 1$. Then $\psi, \psi + 1, \ldots, \psi + p - 1$ are the roots of $x^p - x = \psi^p - \psi \in K$, and the proof can follow §46. This fails if $s = 0$. If $s = 0$ one can try the same argument with θ replaced by θ^2 and θ^3, etc. If this fails for all powers of θ then the $p \times p$ Vandermonde determinant $|\theta_i^j| = 0$, i.e. $\prod(\theta_\mu - \theta_\nu) = 0$, which is impossible.

15. Use the Proposition of §46, Exercise 14, and the fact that qth roots of unity can be constructed using roots of unity for prime exponents less than q.

16. Let $g(X)$ be irreducible over K and let a be a root of g in the splitting field. Then all *distinct* images Sa of a under the Galois group are roots of g ($g(Sa) = Sg(a) = S(0) = 0$) so the product $\bar{g}(X) = \prod(X - Sa)$ over distinct Sa divides $g(X)$. Since $g(X)$ is irreducible and $\bar{g}(X)$ has coefficients in K, g is a constant multiple of $\bar{g}$. Thus the splitting field is a normal extension. Now let K' be the subfield of the splitting field corresponding to the subgroup H of the Galois group G. Suppose H is a normal subgroup. If g is irreducible over K and has a root a in K', and if b is any other root of g then $b = Sa$ for some S. Then for any T in H, $S^{-1}TS = T_0 \in H$ and $Tb = TSa = ST_0a = Sa = b$, so $b \in K'$ and K' is normal. Conversely, suppose K' is a normal extension of K. If $T \in H$, $S \in G$, and $a \in K'$ then $\bar{g}(X) = \prod(X - Sa)$ is irreducible, as above, so $Sa \in K'$. Therefore $TSa = Sa$, $S^{-1}TSa = a$. Since $S^{-1}TS$ leaves all elements of K' invariant, $S^{-1}TS \in H$ and H is normal.

17. By the theorem of the primitive element (Exercises 13, 14, Fourth Set) $K' = K(t)$ where t is a root of an irreducible polynomial with coefficients in K. Since K' is normal, it is a splitting field for this polynomial.

18. Let $b = \prod Sa$. The norm of a product is the product of the norms. The norm of any Sa is the norm of a, as one finds by applying S to the basis used to define the norm. Thus $Nb = (Na)^n$ where n = number of elements in the Galois group = $[K':K]$. On the other hand, $Nb = b^n$ because $b \in K$. Thus Na is b times an nth root of unity. To see that this root is $+1$ apply the argument just used to $\prod (X - Sa)$ in place of $\prod Sa$. This is a polynomial $F(X)$ with coefficients in K. One finds $F(X)^n = [N(X - a)]^n$. By unique factorization of polynomials, $F(X)$ is a constant in K times $N(X - a)$. Since both have leading coefficient 1, the constant is 1. Set $X = 0$.

19. See Exercise 11, Sixth Set. A more direct proof, not using Dedekind's theorem, is not difficult to give using the approach of §44.

List of Exercises

First Exercise Set

1. Quadratic formula (real roots implies discriminant positive) 2. Quadratic formula as algebraic solution 3. $x^3 + 6x = 20$, 4. $x^3 = 3x + 2$ 5. Continuation of 3 6. Elimination of second term of general nth degree equation 7. Solution of quartics 8. Resultant of two cubics 9. Continuation 10. Continuation 11. Lexicographic order 12. Alternate proof of fundamental theorem on symmetric functions 13. Derivation of formulas of §9 14. Alternate proof of a formula of §11 15. Proof of Newton's theorem 16. Derivation of formulas of §9 17. Root cubing 18. Root squaring 19. Computation with symmetric polynomials 20. Symmetric polynomials in terms of sums of powers 21. Continuation 22. Uniqueness in the fundamental theorem 23. Vandermonde determinants 24. Discriminant of a quartic 25. Discriminant of a cubic 26. Continuation 27. Formulas for sums of powers 28. Kronecker's canonical form for polynomials in n variables 29. Supplement to proof of fundamental theorem 30. Cramer's rule.

Second Exercise Set

1. Generators for permutations of three objects 2. Application of fundamental theorem to §14 3. Application to §16 4. Solution of quartics by Lagrange resolvent 5. Application of fundamental theorem to §17 6. Degeneracy in solution of quartic 7. Lagrange resolvent of a quintic 8. Quartics for which Lagrange method fails.

Third Exercise Set

1. Construction of equilateral triangle 2. Construction of regular pentagon 3. 5th roots of unity 4. Primitive nth roots of unity for n a product of relatively prime factors 5. 11th roots of unity 6. pth roots of unity 7. Primitive roots mod p 8. A polynomial in α invariant under $\alpha \rightarrow \alpha^g$ is a rational number (α = primitive pth root of unity, g =

primitive root mod p)   **9.** Intermediate fields between $\mathbb{Q}$ and $\mathbb{Q}(\alpha)$ **10.** 5th roots of unity **11.** 7th roots of unity **12.** 17th roots of unity **13.** Continuation **14.** Construction of square roots   **15.** p^nth roots of unity

Fourth Exercise Set

1. Permutation of variables in polynomials   **2.** Derivatives of polynomials, reduction of solution of equations to equations with simple roots   **3.** A nonzero polynomial over an infinite field has nonzero values   **4.** Product of nonzero polynomials nonzero **5.** Complex numbers   **6.** Division in $\mathbb{Q}(\sqrt{2})$ **7.** $a - b$ a Galois resolvent of a cubic **8.** Galois resolvent of a cubic with square discriminant has degree 3   **9.** Explicit solution of cubic using Galois resolvent   **10.** Continuation   **11.** Continuation   **12.** Continuation **13.** Theorem of primitive element   **14.** Continuation   **15.** Continuation of Exercise 28 of the First Set   **16.** Existence of Galois resolvent, alternate proof **17.** Galois resolvent of quadratic   **18.** Division of polynomials   **19.** Continuation

Fifth Exercise Set

1. Group if and only if closed under composition   **2.** Subgroup gives partition of presentation   **3.** Normal subgroups   **4.** Transitivity and irreducibility   **5.** Galois group of a cubic   **6.** Explicit constructions in the cubic case   **7.** Continuation   **8.** Explicit cubics with cyclic Galois group   **9.** All irreducible factors of $F(X)$ have same degree   **10.** Galois group of $x^p - 1 = 0$ over $\mathbb{Q}$ **11.** $x^4 + 1 = 0$ **12.** $x^8 - 1 = 0$ **13.** Galois' example of a normal subgroup   **14.** Subgroups of cyclic groups are cyclic **15.** Subgroup of index 2 is normal   **16.** A lemma for §42   **17.** A lemma for §41   **18.** Galois group given by a primitive element for the splitting field (not necessarily a Galois resolvent)

Sixth Exercise Set

1. Solution of cubic by radicals using Galois' method   **2.** Of quartic   **3.** Trick for finding discriminant of a quartic   **4.** If one root can be expressed in terms of radicals then all can   **5.** The equation $(x^p - 1)/(x - 1) = 0$ is solvable   **6.** Proposition II **7.** Proposition III   **8.** Liouville on Proposition II   **9.** Proposition IV   **10.** Dedekind's "reciprocity" theorem **11.** Application **12.** Comparison of solutions of cubic **13.** When adjunctions reduce the Galois group

Seventh Exercise Set

1. Unique factorization of polynomials **2.** Corollary 1 of §57 **3.** Irreducible polynomials in two variables are prime   **4.** $K[a, u]$ compared to $K(a)[x]$ **5.** Finding g given $\bar{g}$ (§59)   **6.** A factoring problem   **7.** A norm   **8.** A form of Gauss's lemma **9.** Norms as products of conjugates   **10.** Definition of separability   **11.** Perfect fields (characteristic p)   **12.** Norm in a nonseparable field   **13.** Simple transcendental extensions   **14.** F is a root of $N(X - F)$.

Eighth Exercise Set

1. Basis of a vector space   **2.** Basis of a composite field extension   **3.** A lemma for §64
4. Corollary 2 of §68   **5.** A lemma for §68   **6.** If $x^p - k = 0$ is reducible it has a root
7. Alternate derivation of the Galois group of the general nth degree equation   **8.**
Eisenstein irreducibility criterion   **9.** Fermat's theorem   **10.** Galois theory in characteristic p **11.** Continuation   **12.** Continuation   **13.** Continuation   **14.** Continuation
15. Continuation   **16.** Normal extensions in the "fundamental theorem of Galois
theory"   **17.** continuation   **18.** Norm as a product of conjugates   **19.** Factorization
of irreducible polynomials over normal extensions.

References

[A1]   Aaboe, A., *Episodes from the Early History of Mathematics*, New Mathematical Library, New York, 1964.

[A2]   Abel, N. H., *Oeuvres Complètes de Niels Henrik Abel*, Vol. 1, Grondahl, Christiana, 1881.

[D1]   Dupuy, P., La vie d'Évariste Galois, *Annales de l'École Normale* (3) **XIII** (1896), 197–266.

[E1]   Edwards, H. M., *Fermat's Last Theorem*, Springer-Verlag, New York, 1977.

[E2]   Edwards, H. M., Read the Masters! in *Mathematics Tomorrow*, L. A. Steen, Editor, Springer-Verlag, New York, 1981.

[E3]   Edwards, H. M., *Riemann's Zeta Function*, Academic Press, New York, 1974.

[G1]   Galois, É., *Ecrits et Mémoires Mathématiques d'Évariste Galois*, R. Bourgne and J.-P. Azra, Editors, Gauthier-Villars, Paris, 1962.
        Also *Oeuvres Mathématiques d'Évariste Galois*, Gauthier-Villars, Paris, 1897, and *J. Math. Pures et Appl.* (1) **11** (1846), 381–444.

[G2]   Gauss, C. F., *Disquisitiones Arithmeticae*, Braunschweig, 1801, republished, 1863, as vol. 1 of *Werke*; French transl., *Recherches Arithmétiques*, Paris, 1807, republished Hermann, Paris, 1910; German transl., Springer-Verlag, Berlin, 1889, republished Chelsea, New York, 1965; English transl., Yale, New Haven and London, 1966.

[K1]   Kiernan, B. M., The development of Galois theory from Lagrange to Artin, *Arch. Hist. Exact Sci.*, **8** (1971), 40–154.

[K2]   Kline, M., *Mathematical Thought from Ancient to Modern Times*, Oxford Univ. Press, New York, 1972.

[K3]   Kronecker, L., Beweis, dass für jede Primzahl p die Gleichung $1 + x + x^2 + \cdots + x^{p-1} = 0$ irreductibel ist, *Crelle*, **29** (1845) 280; also, *Werke*, vol. 1, pp. 1–4.

[K4]   Kronecker, L., *Grundzüge einer arithmetischen Theorie der algebraischen Grössen*, Reimer, Berlin, 1882; also *Crelle*, **92** (1882) 1–122; also *Werke*, vol. 2, pp. 237–387.

[L1]   Lagrange, J. L., *Réflexions sur la Résolution Algébrique des Équations*, Nouveaux Mémoires de l'Académie royale des Sciences et Belles-Lettres de Berlin, 1770–1771; also *Oeuvres*, vol. 3, pp. 205–421.

[L2]   Lebesgue, H., L'oeuvre mathématique de Vandermonde, in *Notices d'Histoire des Mathématiques*, Institut de Mathématiques, Université, Geneve, 1958; also in *L'Enseignement Mathématique*, 2nd series, **1** (1955), 203–223.

[M1]   McKay, J. H., Another proof of Cauchy's group theorem, *Amer. Math. Monthly*, **66** (1959), 119.

[N1]   Neugebauer, O., *The Exact Sciences in Antiquity*, Princeton Univ. Press, Princeton, 1952.

[N2]   Newton, I., *Arithmetica Universalis*, 1707; English transl., *Universal Arithmetick*, London, 1728; also in *The Mathematical Works of Isaac Newton*, vol. 2, Johnson Reprint Corp., 1967.

[N3]   Newton, I., *The Mathematical Papers of Isaac Newton*, D. T. Whiteside, Editor, Vol. I, Cambridge Univ. Press, Cambridge, 1967.

[R1]   Rothman, T., Genius and biographers—the fictionalization of Évariste Galois, *Amer. Math. Monthly*, **89** (1982), 84–106.

[S1]   Scharlau, W., Unveröffentlichte algebraische Arbeiten Richard Dedekinds aus seiner Göttinger Zeit 1855–1858, *Arch. Hist. Exact Sci.*, **27** (1982), 335–367.

[T1]   Taton, R., Les rélations d'Évariste Galois avec les mathématiciens de son temps, *Révue d'histoire des sciences et de leurs applications*, **I** (1947), 114–130.

[V1]   Vandermonde, A.-T., Mémoire sur la résolution des équations, Histoire de l'Académie royale des sciences, 1771 (actually 1774); German transl. in *Abhandlungen aus der reinen Mathematik von N. Vandermonde*, Springer, Berlin, 1888.

[W1]   van der Waerden, B. L., On the sources of my book *Moderne Algebra*, *Historia Math.*, **2** (1975), 31–40.

[W2]   van der Waerden, B. L., *Moderne Algebra*, Springer-Verlag, Berlin, 1930–31; later editions 1937, 1950, 1955, 1960, 1964, 1966, 1971; English transl., Ungar, New York, 1949, 1953, 1970.

[W3]   Weber, H., *Lehrbuch der Algebra*, Vieweg, Braunschweig, 1895, 1896; reprint Chelsea, New York, 1962, 1979.

Index

Graduate Texts in Mathematics

continued from page ii